The Michigan Historical Reprint Series

Reprints from the collection of the University of Michigan University Library

This volume is produced from digital images created through the University of Michigan University Library's preservation reformatting program. The Library seeks to preserve the intellectual content of items in a manner that facilitates and promotes a variety of uses. The digital reformatting process results in an electronic version of the text that can both be accessed online and used to create new print copies. This book and thousands of others can be found in the digital collections of the University of Michigan Library. The University Library also understands and values the utility of print, and makes reprints available through its Scholarly Publishing Office.

For access to the University of Michigan Library's digital collections, please see http://www.lib.umich.edu.

The Scholarly Publishing Office seeks to disseminate high-quality, cost-effective scholarly content through both print and electronic publishing. Information about the Scholarly Publishing Office can be found at http://spo.umdl.umich.edu.

The Scholarly Publishing Office
the University of Michigan
University Library

DISCOURSES

ON THE

CHRISTIAN REVELATION

VIEWED IN CONNECTION WITH THE

MODERN ASTRONOMY.

TO WHICH ARE ADDED

DISCOURSES

ILLUSTRATIVE OF THE CONNECTION BETWEEN THEOLOGY AND GENERAL SCIENCE.

BY

THOMAS CHALMERS, D.D. & LL.D.

PROFESSOR OF THEOLOGY IN THE UNIVERSITY OF EDINBURGH
AND CORRESPONDING MEMBER OF THE ROYAL INSTITUTE OF FRANCE.

NEW YORK:
ROBERT CARTER & BROTHERS
No. 285 BROADWAY.

1855.

PREFACE.

THE astronomical objection against the truth of the Gospel, does not occupy a very prominent place in any of our Treatises of Infidelity. It is often, however, met with in conversation—and we have known it to be the cause of serious perplexity and alarm in minds anxious for the solid establishment of their religious faith.

There is an imposing splendour in the science of Astronomy; and it is not to be wondered at, if the light it throws, or appears to throw, over other tracks of speculation than those which are properly its own, should at times dazzle and mislead an inquirer. On this account, we think it were a service to what we deem a true and a righteous cause, could we succeed in dissipating this illusion, and in stripping Infidelity of those pretensions to enlargement, and to a certain air of philosophical greatness, by which it has often become so destructively alluring to the young, and the ardent, and the ambitious.

In my first Discourse, I have attempted a sketch of the Modern Astronomy—nor have I wished to throw any disguise over that comparative littleness which belongs to our planet, and which gives to the argument of Freethinkers all its plausibility.

This argument involves in it an assertion and an inference. The assertion is, that Christianity is a religion which professes to be designed for the single benefit of our world; and the inference is, that God cannot be the author of this religion, for He would not lavish on so insignificant a field, such peculiar and such distinguishing attentions, as are ascribed to Him in the Old and New Testament.

Christianity makes no such profession. That it is designed for the single benefit of our world is altogether a presumption of the Infidel himself—and feeling that this is not the only example of temerity which can be charged on the enemies of our faith, I have allotted my second Discourse to the attempt of demonstrating the utter repugnance of such a spirit with the cautious and enlightened philosophy of modern times.

In the course of this Sermon I have offered a tribute of acknowledgment to the theology of Sir Isaac Newton; and in such terms, as if not farther explained, may be liable to misconstruction. The grand circumstance of applause in the character of this great man, is, that unseduced by all the magnificence of his own discoveries, he had a solidity of mind which could resist their fascination, and keep him in steady attachment to that Book, whose general evidences stamped upon it the impress of a real communication from Heaven. This was the sole attribute of his theology which I had in my eye when I presumed to eulogize it. I do not think, that, amid the distraction and the engross ment of his other pursuits, he has at all times succeeded in his interpretation of the Book; else he would never, in my apprehension, have abetted the leading doctrine of a sect or a system, which has now nearly dwindled away from public observation.

In my third Discourse I am silent as to the assertion, and attempt to combat the inference that is founded on it. I insist, that upon all the analogies of nature and of providence, we can lay no limit on the condescension of God, or on the multiplicity of his regards even to the very humblest departments of creation; and that it is not for us, who see the evidences of divine wisdom and care spread in such exhaustless profusion around us, to say, that the Deity would not lavish all the wealth of His wondrous attributes on the salvation even of our solitary species.

At this point of the argument, I trust that the intelligent reader may be enabled to perceive, in the adversaries of the Gospel, a twofold dereliction from the maxims of the Baconian philosophy: that, in the first instance, the assertion which forms the groundwork of their argument, is gratuitously fetched out of an unknown region, where they are utterly abandoned by the light of experience; and that, in the second instance, the inference they urge from it is, in the face of manifold and undeniable truths, all lying within the safe and accessible field of human observation.

In my subsequent Discourses, I proceed to the informations of the Record. The Infidel objection drawn from Astronomy, may be considered as by this time disposed of; and if we have succeeded in clearing it away, so as to deliver the Christian testimony from all discredit upon this ground, then may we submit, on the strength of other evidences, to be guided by its information. We shall thus learn, that Christianity has a far more extensive bearing on the other orders of creation, than the Infidel is disposed to allow; and, whether he will own the authority of this information or not, he will at least be forced to admit, that the subject-matter of the Bible itself is not chargeable with that objection which he has attempted to fasten upon it.

Thus, had my only object been the refutation of the Infidel argument, I might have spared the last Discourses of the Series altogether. But the tracks of Scriptural information to which they directed me, I considered as worthy of prosecution on their own account—and I do think, that much may be gathered from these less observed portions of the field of revelation, to cheer, and to elevate, and to guide the believer.

But in the management of such a discussion as this, though for a great degree of this effect it would require to be conducted in a far higher style than I am able to sustain, the taste of the human mind may be regaled, and its understanding put into a state of the most agreeable exercise. Now, this is quite distinct from the conscience being made to feel the force of a personal application; nor could I either bring this argument to its close in the pulpit, or offer it to the general notice of the world, without adverting, in the last Discourse, to a delusion, which, I fear, is carrying forward thousands, and tens of thousands, to an undone eternity.

I have closed the Series with an Appendix of Scriptural Authorities. I found that I could not easily interweave them in the texture of the Work, and have, therefore, thought fit to present them in a separate form. I look for a twofold benefit from this exhibition—first, to those more general readers, who are ignorant of the Scriptures, and of the richness and variety which abound in them—and, secondly, to those narrow and intolerant professors, who take an alarm at the very sound and semblance of philosophy; and feel as if there was an utterly irreconcilable antipathy between its lessons on the one hand, and the soundness and piety of the Bible on the other. It were well, I conceive, for our cause, that the latter could become a little more indulgent on this subject; that they gave up a portion of those ancient and hereditary prepossessions, which go so far to cramp and to enthral them; that they would suffer theology to take that wide range of argument and of illustration which belongs to her; and that, less sensitively jealous of any desecration being brought upon the Sabbath or the pulpit, they would suffer her freely to announce all those truths, which either serve to protect Christianity from the contempt of science, or to protect the teachers of Christianity from those invasions, which are practised both on the sacredness of the office, and on the solitude of its devotional and intellectual labours.

To these Astronomical Discourses, I have added some others, illustrative of the connexion between

Theology and General Science. The argument on which we have ventured in one of these Discourses, and by which we attempt to reconcile the efficacy of prayer with the constancy of visible nature, was called forth in opposition to the contemptuous treatment, which certain members of the British Senate thought fit to bestow on the proposal for a National Fast, at a time when the fearful epidemic of cholera had broke forth in various parts of the country.

CONTENTS.

DISCOURSE III.

DISCOURSE IV.

DISCOURSE V.

DISCOURSE VI.

DISCOURSE VII.

DISCOURSE I.

A SKETCH OF THE MODERN ASTRONOMY.

"When I consider thy heavens, the work of thy fingers, the moon and the stars, which thou hast ordained; What is man, that thou art mindful of him? and the son of man, that thou visitest him?"—PSALM viii. 3, 4.

IN the reasonings of the Apostle Paul, we cannot fail to observe, how studiously he accommodates his arguments to the pursuits or principles or prejudices of the people whom he was addressing. He often made a favourite opinion of their own the starting point of his explanation; and, educing a dexterous but irresistible train of argument from some principle upon which each of the parties had a common understanding, it was his practice to force them out of all their opposition, by a weapon of their own choosing,—nor did he scruple to avail himself of a Jewish peculiarity, or a heathen superstition, or a quotation from Greek poetry, by which he might gain the attention of those whom he laboured to convince, and by the skilful application of which he might "shut them up unto the faith."

Now, when Paul was thus addressing one class of an assembly, or congregation, another class might, for the time, have been shut out of all direct benefit and application from his arguments. When he wrote an Epistle to a mixed assembly of Christianized Jews and Gentiles, he had often to direct such a process of argument to the former, as the latter would neither require nor comprehend. Now, what should have been the conduct of the Gentiles at the reading of that part of the Epistle which bore almost an exclusive reference to the Jews? Should it be impatience at the hearing of something for which they had no relish or understanding? Should it be a fretful disappointment, because every thing that was said, was not said for their edification? Should it be angry discontent with the Apostle, because, leaving them in the dark, he had brought forward nothing for them, through the whole extent of so many successive chapters? Some of them may have felt in this way; but surely it would have been vastly more Christian to have sat with meek and unfeigned patience, and to have rejoiced that the great Apostle had undertaken the management of those obstinate prejudices, which kept back so many human beings from the participation of the Gospel. And should Paul have had reason to rejoice, that, by the success of his arguments, he had reconciled one or any number of Jews to Christianity, then it was the part of these Gentiles, though receiving no direct or personal benefit from the arguments, to have blessed God, and rejoiced along with him.

Conceive that Paul were at this moment alive, and zealously engaged in the work of pressing the Christian religion on the acceptance of the various classes of society. Should he not still have acted on the principle of being all things to all men? Should he not have accommodated his discussion to the prevailing taste, and literature, and philosophy of the times? Should he not have closed with the people, whom he was addressing, on some favourite principle of their own; and, in the prosecution of this principle, might he not have got completely beyond the comprehension of a numerous class of zealous, humble, and devoted Christians? Now, the question is not, how these would conduct themselves in such circumstances? but, how should they do it? Would it be right in them to sit with impatience, because the argument of the Apostle contained in it nothing in the way of comfort or edification to themselves? Should not the benevolence of the Gospel give a different direction to their feelings? And, instead of that narrow, exclusive, and monopolising spirit, which I fear is too characteristic of the more declared professors of the truth as it is in Jesus, ought they not to be patient, and to rejoice, when to philosophers, and to men of literary accomplishment, and to those who have the direction of the public taste among the upper walks of society, such arguments are addressed as may bring home to their acceptance also, "the words of this life?" It is under the impulse of these considerations that I have, with some hesitation, prevailed upon myself to attempt an argument, which I think fitted to soften and subdue those prejudices which lie at the bottom of what may be called the infidelity of natural science; if possible to bring over to the humility of the Gospel, those who expatiate with delight on the wonders and the sublimities of creation; and to convince them, that a loftier wisdom still than that even of their high and honourable acquirements, is the wisdom of him who is resolved to know nothing but Jesus Christ and Him crucified.

It is truly a most Christian exercise to extract a sentiment of piety from the works and the appearances of nature. It has the authority of the Sacred Writers upon its side, and even our Saviour himself gives it the weight and the solemnity of his example. "Behold the lilies of the field; they toil not, neither do they spin, yet your heavenly Father careth for them." He expatiates on the beauty of a single flower, and draws from it the delightful argument of confidence in God. He gives us to see that taste may be combined with piety, and that the same heart may be occupied with all that is serious in the contemplations of religion, and be at the same time alive to the charms and the loveliness of nature.

The Psalmist takes a still loftier flight. He leaves the world, and lifts his imagination to that mighty expanse which spreads above it and around it. He wings his way through space, and wanders in thought over its immeasurable regions. Instead of a dark and unpeopled solitude, he sees it crowded with splendour, and filled with the energy of the Divine presence. Creation rises in its immensity before him; and the world, with all which it inherits, shrinks into littleness at a contemplation so vast and so overpowering. He wonders that he is not overlooked amid the grandeur and the variety which are on every side of him; and passing upward from the majesty of nature to the majesty of nature's Architect, he exclaims, "What is man, that thou art mindful of him; or the son of man, that thou shouldest deign to visit him?"

It is not for us to say, whether inspiration revealed to the Psalmist the wonders of the modern astronomy. But even though the mind be a perfect stranger to the science of these enlightened times, the heavens present a great and an elevating spectacle—an immense concave reposing upon the circular boundary of the world, and the innumerable lights which are suspended from on high, moving with solemn regularity along its surface. It seems to have been at night that the piety of the Psalmist was awakened by this contemplation, when the moon and the stars were visible, and not when the sun had risen in his strength, and thrown a splendour around him, which bore down and eclipsed all the lesser glories of the firmament. And there is much in the scenery of a nocturnal sky, to lift the soul to pious contemplation. That moon, and these stars, what are they? They are detached from the world, and they lift us above it. We feel withdrawn from the earth, and rise in lofty abstraction from this little theatre of human passions and human anxieties. The mind abandons itself to reverie, and is transferred in the ecstasy of its thoughts, to distant and unexplored regions. It sees nature in the simplicity of her great elements, and it sees the God of nature invested with the high attributes of wisdom and majesty.

But what can these lights be? The curiosity of the human mind is insatiable; and the mechanism of these wonderful heavens has, in all ages, been its subject and its employment. It has been reserved for these latter times, to resolve this great and interesting question. The sublimest powers of philosophy have been called to the exercise, and astronomy may now be looked upon as the most certain and best established of the sciences.

We all know that every visible object appears less in magnitude as it recedes from the eye. The lofty vessel, as it retires from the coast, shrinks into littleness, and at last appears in the form of a small speck on the verge of the horizon. The eagle, with its expanded wings, is a noble object; but when it takes its flight into the upper regions of the air, it becomes less to the eye, and is seen like a dark spot upon the vault of heaven. The same is true of all magnitude. The heavenly bodies appear small to the eye of an inhabitant of this earth, only from the immensity of their distance. When we talk of hundreds of millions of miles, it is not to be listened to as incredible. For remember that we are talking of those bodies which are scattered over the immensity of space, and that space knows no termination. The conception is great and difficult, but the truth is unquestionable. By a process of measurement which it is unnecessary at present to explain, we have ascertained first the distance, and then the magnitude of some of those bodies which roll in the firmament; that the sun which presents itself to the eye under so diminutive a form, is really a globe, exceeding, by many thousands of times, the dimensions of the earth which we inhabit; that the moon itself has the magnitude of a world; and that even a few of those stars, which appear like so many lucid points to the unassisted eye of the observer, expand into large circles upon the application of the telescope, and are some of them much larger than the ball which we tread upon, and to which we proudly apply the denomination of the universe.

Now, what is the fair and obvious presumption? The world in which we live, is a round ball of a determined magnitude, and occupies its own place in the firmament. But when we explore the unlimited tracts of that space, which is every where around us, we meet with other balls of equal or superior magnitude, and from which our earth would either be invisible, or appear as small as any of those twinkling stars which are seen on the canopy of heaven. Why then suppose that this little spot, little at least in the immensity which surrounds it, should be the exclusive abode of life and of intelligence? What reason to think that those mightier globes which roll in other parts of creation, and which we have discovered to be worlds in magnitude, are not also worlds in use and in dignity? Why should we think that the great Architect of nature, supreme in wisdom, as He is in power, would call these stately mansions into existence and leave them unoccupied? When we cast our eye over the broad sea, and look at the country on the other side, we see nothing but the blue land stretching obscurely over the distant horizon. We are too far away to perceive the richness of its scenery, or to hear the sound of its population Why not extend this principle to the still more distant parts of the universe? What though, from this remote point of observation, we can see nothing but the naked roundness of yon planetary orbs? Are we therefore to say, that they are so many vast and unpeopled solitudes; that desolation reigns in every part of the universe but ours; that the whole energy of the divine attributes is expended on one insignificant corner of these mighty works; and that to this earth alone belongs the bloom of vegetation, or the blessedness of life, or the dignity of rational and immortal existence?

But this is not all. We have something more than the mere magnitude of the planets to allege in favour of the idea that they are inhabited. We know that this earth turns round upon itself; and we observe that all those celestial bodies, which are accessible to such an observation, have the same movement. We know that the earth performs a yearly revolution round the sun; and we can detect, in all the planets which compose our system, a revolution of the same kind, and under the same circumstances. They have the same succession of day and night. They have the same agreeable vicissitude of the seasons. To them light and darkness succeed each other; and the gaiety of summer is followed by the dreariness of winter. To each of them the heavens present as varied and magnificent a spectacle; and this earth, the encompassing of which would require the labour of years from one of its puny inhabitants, is but one of the lesser lights which sparkle in their firmament. To them, as well as to us, has God divided the light from the darkness, and he has called the light day, and the darkness he has called night. He has said, let there be lights in the firmament of their heaven, to divide the day from the night; and let them be for signs, and for seasons, and for days, and for years; and let them be for lights in the firmament of heaven, to give light upon their earth; and it was so. And God has also made to them great lights. To all of them he has given the sun to rule the day; and to many of them has he given moons to rule the night. To them he has made the stars also. And God has set them in the firmament of heaven, to give light upon their earth; and to rule over the day, and over the night, and to divide the light from the darkness; and God has seen that it was good.

In all these greater arrangements of divine wisdom, we can see that God has done the same things for the accommodation of the planets that he has done for the earth which we inhabit. And shall we say, that the resemblance stops here, because we are not in a situation to observe it? Shall we say, that this scene of magnificence has been called into being merely for the amusement of a few astronomers? Shall we measure the councils of heaven by the narrow impotence of the human faculties? or conceive, that silence and solitude reign throughout the mighty empire of nature; that the greater part of creation is an empty parade; and that not a worshipper of the Divinity is to be found through the wide extent of yon vast and immeasurable regions?

It lends a delightful confirmation to the argument, when, from the growing perfection of our instruments, we can discover a new point of resemblance between our Earth and the other bodies of the planetary system. It is now ascertained, not merely that all of them have their day and night, and that all of them have their vicissitudes of seasons, and that some of them have their moons to rule their night and alleviate the darkness of it; —we can see of one, that its surface rises into inequalities, that it swells into mountains and stretches into valleys; of another, that it is surrounded by an atmosphere which may support the respiration of animals; of a third, that clouds are formed and suspended over it, which may minister to it all the bloom and luxuriance of vegetation; and of a fourth, that a white colour spreads over its northern regions, as its winter advances, and that, on the approach of summer, this whiteness is dissipated—giving room to suppose, that the element of water abounds in it, that it rises by evaporation into its atmosphere, that it freezes upon the application of cold, that it is precipitated in the form of snow, that it covers the ground with a fleecy mantle, which melts away from the heat of a more vertical sun; and that other worlds bear a resemblance to our own, in the same yearly round of beneficent and interesting changes.

Who shall assign a limit to the discoveries of future ages? Who can prescribe to science her boundaries, or restrain the active and insatiable curiosity of man within the circle of his present acquirements? We may guess with plausibility what we cannot anticipate with confidence. The day may yet be coming, when our instruments of observation shall be inconceivably more powerful

They may ascertain still more decisive points of resemblance. They may resolve the same question by the evidence of sense, which is now so abundantly convincing by the evidence of analogy. They may lay open to us the unquestionable vestiges of art, and industry, and intelligence. We may see summer throwing its green mantle over these mighty tracts, and we may see them left naked and colourless after the flush of vegetation has disappeared. In the progress of years or of centuries, we may trace the hand of cultivation spreading a new aspect over some portion of a planetary surface. Perhaps some large city, the metropolis of a mighty empire, may expand into a visible spot by the powers of some future telescope. Perhaps the glass of some observer, in a distant age, may enable him to construct the map of another world, and to lay down the surface of it in all its minute and topical varieties. But there is no end of conjecture; and to the men of other times we leave the full assurance of what we can assert with the highest probability, that yon planetary orbs are so many worlds, that they teem with life, and that the mighty Being who presides in high authority over this scene of grandeur and astonishment, has there planted the worshippers of His glory.

Did the discoveries of science stop here, we have enough to justify the exclamation of the Psalmist, "What is man, that thou art mindful of him; or the son of man, that thou shouldest deign to visit him?" They widen the empire of creation far beyond the limits which were formerly assigned to it. They give us to see that yon sun, throned in the centre of his planetary system, gives light, and warmth, and the vicissitude of seasons, to an extent of surface several hundreds of times greater than that of the earth which we inhabit. They lay open to us a number of worlds, rolling in their respective circles around this vast luminary—and prove, that the ball which we tread upon, with all its mighty burden of oceans and continents, instead of being distinguished from the others, is among the least of them; and, from some of the more distant planets, would not occupy a visible point in the concave of their firmament. They let us know, that though this mighty earth, with all its myriads of people, were to sink into annihilation, there are some worlds where an event so awful to us would be unnoticed and unknown, and others where it would be nothing more than the disappearance of a little star which had ceased from its twinkling. We should feel a sentiment of modesty at this just but humiliating representation. We should learn not to look on our earth as the universe of God, but one paltry and insignificant portion of it; that it is only one of the many mansions which the Supreme Being has created for the accommodation of His worshippers, and only one of the many worlds rolling in that flood of light which the sun pours around him to the outer limits of the planetary system.

But is there nothing beyond these limits? The planetary system has its boundary, but space has none; and if we wing our fancy there, do we only travel through dark and unoccupied regions? There are only five, or at most six, of the planetary orbs visible to the naked eye. What, then, is that multitude of other lights which sparkle in our firmament, and fill the whole concave of heaven with innumerable splendours? The planets are all attached to the sun; and, in circling around him, they do homage to that influence which binds them to perpetual attendance on this great luminary. But the other stars do not own his dominion. They do not circle around him. To all common observation, they remain immoveable; and each, like the independent sovereign of his own territory, appears to occupy the same inflexible position in the regions of immensity. What can we make of them? Shall we take our adventurous flight to explore these dark and untravelled dominions? What mean these innumerable fires lighted up in distant parts of the universe? Are they only made to shed a feeble glimmering over this little spot in the kingdom of nature? or do they serve a purpose worthier of themselves, to light up other worlds, and give animation to other systems?

The first thing which strikes a scientific observer of the fixed stars, is their immeasurable distance. If the whole planetary system were lighted up into a globe of fire, it would exceed, by many millions of times, the magnitude of this world, and yet only appear a small lucid point from the nearest of them. If a body were projected from the sun with the velocity of a cannon-ball, it would take hundreds of thousands of years before it described that mighty interval which separates the nearest of the fixed stars from our sun and from our system. If this earth, which moves at more than the inconceivable velocity of a million and a half miles a-day, were to be hurried from its orbit, and to take the same rapid flight over this immense tract, it would not have arrived at the termination of its journey, after taking all the time which has elapsed since the creation of the world. These are great numbers, and great calculations; and the mind feels its own impotency in attempting to grasp them. We can state them in words. We can exhibit them in figures. We can demonstrate them by the powers of a most rigid and infallible geometry. But no human fancy can summon up a lively or an adequate conception—can roam in its ideal flight over this immeasurable largeness—can take in this mighty space in all its grandeur, and in all its immensity—can sweep the outer boundaries of such a creation—or lift itself up to the majesty of that great and invisible arm on which all is suspended.

But what can those stars be which are seated so far beyond the limits of our planetary system? They must be masses of immense magnitude, or they could not be seen at the distance of place which they occupy. The light which they give must proceed from themselves, for the feeble reflection of light from some other quarter, would not carry through such mighty tracts to the eye of an observer. A body may be visible in two ways. It may be visible from its own light, as the flame of a candle, or the brightness of a fire, or the brilliancy of yonder glorious sun, which lightens all below, and is the lamp of the world. Or it may be visible from the light which falls upon it, as the body which receives its light from a taper—or the whole assemblage of objects on the surface of the earth, which appear only when the light of day rests upon them—or the moon, which, in that part of it that is towards the sun, gives out a silvery whiteness to the eye of the observer, while the other part forms a black and invisible space in the firmament—or as the planets, which shine only because the sun shines upon them, and which, each of them, present the appearance of a dark spot on the side that is turned away from it. Now apply this question to the fixed stars. Are they luminous of themselves, or do they derive their light from the sun, like the bodies of our planetary system? Think of their immense distance, and the solution of this question becomes evident. The sun, like any other body, must dwindle into a less apparent magnitude as you retire from it. At the prodigious distance even of the very nearest of the fixed stars, it must have shrunk into a small indivisible point. In short, it must have become a star itself, and could shed no more light than a single individual of those glimmering myriads, the whole assemblage of which cannot dissipate and can scarcely alleviate the midnight darkness of our world. These stars are visible to us, not because the sun shines upon them, but because they shine of themselves, because they are so many luminous bodies scattered over the tracts of immensity—in a word, because they are so many suns, each throned in the centre of his own dominions, and pouring a flood of light over his own portion of these unlimitable regions.

At such an immense distance for observation, it is not to be supposed, that we can collect many points of resemblance between the fixed stars, and the solar star which forms the centre of our plane tary system. There is one point of resemblance, however, which has not escaped the penetration of our astronomers. We know that our sun turns round upon himself, in a regular period of time. We also know that there are dark spots scattered over his surface, which, though invisible to the naked eye, are perfectly noticeable by our instruments. If these spots existed in greater quantity upon one side than upon another, it would have the general effect of making that side darker; and the revolution of the sun must, in such a case, give us a brighter and a fainter side, by regular alternations. Now, there are some of the fixed stars which present this appearance. They present us with periodical variations of light. From the splendour of a star of the first or second magnitude, they fade away into some of the inferior magnitudes—and one, by becoming invisible, might give reason to apprehend that we had lost him altogether—but we can still recognize him by the telescope, till at length he reappears in his own place, and, after a regular lapse of so many days and hours, recovers his original brightness. Now, the fair inference from this is, that the fixed stars, as they resemble our sun in being so many luminous masses of immense magnitude, they resemble him in this also, that each of them turns round upon his own axis; so that if any of them should have an inequality in the brightness of their sides, this revolution is rendered evident, by the regular variations in the degree of light which it undergoes.

Shall we say, then, of these vast luminaries, that they were created in vain? Were they called into existence for no other purpose than to throw a tide of useless splendour over the solitudes of immensity? Our sun is only one of these luminaries, and we know that he has worlds in his train. Why should we strip the rest of this princely attendance? Why may not each of them be the centre of his own system, and give light to his own worlds? It is true that we see them not; but could the eye of man take its flight into those distant regions, it would lose sight of our little world before it reached the outer limits of our system—the greater planets would disappear in their turn—before it had described a small portion of that abyss which separates us from the fixed stars, the sun would decline into a little spot, and all its splendid retinue of worlds be lost in the obscurity of distance—he would at last shrink into a small indivisible atom, and all that could be seen of this magnificent system, would be reduced to the glimmering of a little star. Why resist any longer the grand and interesting conclusion? Each of these stars may be the token of a system as vast and as splendid as the one which we inhabit. Worlds roll in these distant regions; and these worlds must be the mansions of life and of intelligence. In yon gilded canopy of heaven, we see the broad aspect of the universe, where each shining point presents us with a sun, and each sun with a system of worlds—where the Divinity reigns in all the grandeur of His attributes—where He peoples immensity with His wonders; and travels in the greatness of His strength through the dominions of one vast and unlimited monarchy.

The contemplation has no limits. If we ask the number of suns and of systems, the unassisted eye of man can take in a thousand, and the best telescope which the genius of man has constructed can take in eighty millions. But why subject the dominions of the universe to the eye of man, or to the powers of his genius? Fancy may take its flight far beyond the ken of eye or of telescope. It may expatiate in the outer regions of all that is visible—and shall we have the boldness to say, that there is nothing there? that the wonders of the Almighty are at an end, because we can no longer trace His footsteps? that his omnipotence is exhausted, because human art can no longer follow Him? that the creative energy of God has sunk into repose, because the imagination is enfeebled by the magnitude of its efforts, and can keep no longer on the wing through those mighty tracts, which shoot far beyond what eye hath seen, or the heart of man hath conceived—which sweep endlessly along, and merge into an awful and mysterious infinity?

Before bringing to a close this rapid and imperfect sketch of our modern astronomy, it may be right to advert to two points of interesting speculation, both of which serve to magnify our conceptions of the universe, and, of course, to give us a more affecting sense of the comparative insignificance of this our world. The first is suggested by the consideration, that if a body be struck in the direction of its centre, it obtains, from this impulse, a progressive motion, but without any movement of revolution being at the same time impressed upon it. It simply goes forward, but does not turn round upon itself. But, again, should the stroke not be in the direction of the centre—should the line which joins the point of percussion to the centre, make an angle with that line in which the impulse was communicated, then the body is both made to go forward in space, and also to wheel upon its axis. In this way, each of our planets may have had its compound motion communicated to it by one single impulse; and, on the other hand, if ever the rotatory motion be communicated by one blow, then the progressive motion must go along with it. In order to have the first motion without the second, there must be a two-fold force applied to the body in opposite directions. It must be set a-going in the same way as a spinning-top, so as to revolve about an axis, and to keep unchanged its situation in space. The planets have both motions; and, therefore, may have received them by one and the same impulse. The sun, we are certain, has one of these motions. He has a movement of revolution. If spun round his axis by two opposite forces, one on each side of him, he may have this movement, and retain an inflexible position in space. But if this movement was given him by one stroke, he must have a progressive motion along with a whirling motion; or, in other words, he is moving forward; he is describing a tract in space; and, in so doing, he carries all his planets and all their secondaries along with him.

But, at this stage of the argument, the matter only remains a conjectural point of speculation. The sun may have had his rotation impressed upon him by a spinning impulse; or, without recurring to secondary causes at all, this movement may be coeval with his being, and he may have derived both the one and the other from an immediate fiat of the Creator. But there is an actually observed phenomenon of the heavens, which advances the conjecture into a probability. In the course of ages, the stars in one quarter of the celestial sphere are apparently receding from each other; and, in the opposite quarter, they are apparently drawing nearer to each other. If the sun be approaching the former quarter, and receding from the latter, this phenomenon admits of an easy explanation; and we are furnished with a magnificent step in the scale of the Creator's workmanship. In the same manner as the planets, with their satellites, revolve round the sun, may the sun, with all his tributaries, be moving, in common with other stars, around some distant centre, from which there emanates an influence to bind and to subordinate them all. They may be kept from approaching each other, by a centrifugal force; without which, the laws of attraction might consolidate, into one stupendous mass, all the distinct globes of which the universe is composed. Our sun may, therefore, be only one member of a higher family—taking his part, along with millions of others, in some loftier system of mechanism, by which they are all subjected to one law, and to one arrangement—describing the sweep of such an orbit in space, and completing the mighty revolution in such a period of time, as to reduce our planetary seasons, and our planetary movements, to a very humble and fractionary rank in the scale of a higher astronomy. There is room for all this in immensity; and there is even argument for all this, in the records of actual observation; and, from the whole of this speculation, do we gather a new emphasis to the lesson, how minute is the place, and how secondary is the importance of our world, amid the glories of such a surrounding magnificence.

But there is still another very interesting tract of speculation, which has been opened up to us by the more recent observations of astronomy. What we allude to, is the discovery of the *nebulæ*. We allow that it is but a dim and indistinct light which this discovery has thrown upon the structure of the universe; but still it has spread before the eye of the mind a field of very wide and lofty contemplation. Anterior to this discovery, the universe might appear to have been composed of an indefinite number of suns, about equi-distant from each other, uniformly scattered over space, and each encompassed by such a planetary attendance as takes place in our own system. But, we have now reason to think, that instead of lying uniformly, and in a state of equi-distance from each other, they are arranged into distinct clusters—that, in the same manner as the distance of the nearest fixed stars so inconceivably superior to that of our planets from each other, marks the separation of the solar systems, so the distance of two contiguous clusters may be so inconceivably superior to the reciprocal distance of those fixed stars which belong to the same cluster, as to mark an equally distinct separation of the clusters, and to constitute each of them an individual member of some higher and more extended arrangement. This carries us upwards through another ascending step in the scale of magnificence, and there leaves us in the uncertainty, whether even here the wonderful progression is ended; and, at all events, fixes the assured conclusion in our minds, that, to an eye which could spread itself over the whole, the mansion which accommodates our species might be so very small as to lie wrapped in microscopical concealment; and, in reference to the only Being who possesses this universal eye, well might we say, "What is man, that thou art mindful of him; or the son of man, that thou shouldest deign to visit him?"

And, after all, though it be a mighty and difficult conception, yet who can question it? What is seen may be nothing to what is unseen; for what is seen is limited by the range of our instruments. What is unseen has no limit; and, though all which the eye of man can take in, or his fancy can grasp, were swept away, there might still remain as ample a field, over which the Divinity may expatiate, and which He may have peopled with innumerable worlds. If the whole visible creation were to disappear, it would leave a solitude behind it—but to the Infinite Mind, that can take in the whole system of nature, this solitude might be nothing; a small unoccupied point in that immensity which surrounds it, and which he may have filled with the wonders of his omnipotence. Though this earth were to be burned up, though the trumpet of its dissolution were sounded, though yon sky were to pass away as a scroll, and every visible glory, which the finger of the Divinity has inscribed on it, were to be put out for ever—an event, so awful to us, and to every world in our vicinity, by which so many suns would be extinguished, and so many varied scenes of life and of population would rush into forgetfulness—what is it in the high scale of the Almighty's workmanship? a mere shred, which, though scattered into nothing, would leave the universe of God one entire scene of greatness and of majesty. Though this earth, and these heavens, were to disappear, there are other worlds which roll afar; the light of other suns shines upon them; and the sky which mantles them, is garnished with other stars. Is it presumption to say, that the moral world extends to these distant and unknown regions? that they are occupied with people? that the charities of home and of neighbourhood flourish there? that the praises of God are there lifted up, and his goodness rejoiced in? that piety has there its temples and its offerings? and the richness of the divine attributes is there felt and admired by intelligent worshippers?

And what is this world in the immensity which teems with them—and what are they who occupy it? The universe at large would suffer as little, in its splendour and variety, by the destruction of our planet, as the verdure and sublime magnitude of a forest would suffer by the fall of a single leaf. The leaf quivers on the branch which supports it. It lies at the mercy of the slightest accident. A breath of wind tears it from its stem, and it lights on the stream of water which passes underneath. In a moment of time, the life which we know, by the microscope, it teems with, is extinguished; and an occurrence so insignificant in the eye of man, and on the scale of his observation, carries in it, to the myriads which people this little leaf, an event as terrible and as decisive as the destruction of a world. Now, on the grand scale of the universe, we, the occupiers of this ball, which performs its little round among the suns and the systems that astronomy has unfolded—we may feel the same littleness, and the same insecurity. We differ from the leaf only in this circumstance, that it would require the operation of greater elements to destroy us. But these elements exist. The fire which rages within, may lift its devouring energy to the surface of our planet, and transform it into one wide and wasting volcano. The sudden formation of elastic matter in the bowels of the earth —and it lies within the agency of known substances to accomplish this—may explode it into fragments. The exhalation of noxious air from below, may impart a virulence to the air that is around us; it may affect the delicate proportion of its ingredients; and the whole of animated nature may wither and die under the malignity of a tainted atmosphere. A blazing comet may cross this fated planet in its orbit, and realize all the terrors which superstition has conceived of it. We cannot anticipate with precision the consequences of an event which every astronomer must know to lie within the limits of chance and probability. It may hurry our globe towards the sun—or drag it to the outer regions of the planetary system—or give it a new axis of revolution: and the effect, which I shall simply announce, without explaining it, would be to change the place of the ocean, and bring another mighty flood upon our islands and continents. These are changes which may happen in a single instant of time, and against which nothing known in the present system of things provides us with any security. They might not annihilate the earth, but they would unpeople it; and we who tread its surface with such firm and assured footsteps, are at the mercy of devouring elements, which, if let loose upon us by the hand of the Almighty, would spread solitude, and silence, and death, over the dominions of the world.

Now, it is this littleness, and this insecurity, which make the protection of the Almighty so dear to us, and bring, with such emphasis, to every pious bosom, the holy lessons of humility and gratitude. The God who sitteth above, and presides in high authority over all worlds, is mindful of man; and though at this moment His energy is felt in the remotest provinces of creation, we may feel the same security in His providence, as if we were the objects of His undivided care. It is not for us to bring our minds up to this mysterious agency. But such is the incomprehensible fact, that the same Being, whose eye is abroad over the whole universe, gives vegetation to every blade of grass, and motion to every particle of blood which circulates through the veins of the minutest animal; that, though His mind takes into its comprehensive grasp, immensity and all its wonders, I am as much known to Him as if I were the single object of His attention; that He marks all my thoughts; that He gives birth to every feeling and every movement within me; and that, with an exercise of power which I can neither describe nor comprehend, the same God who sits in the highest heaven, and reigns over the glories of the firmament, is at my right hand, to give me every breath which I draw, and every comfort which I enjoy.

But this very reflection has been appropriated to the use of Infidelity, and the very language of the text has been made to bear an application of hostility to the faith. "What is man, that God should be mindful of him; or the son of man, that he should deign to visit him?" Is it likely, says the Infidel, that God would send his eternal Son, to die for the puny occupiers of so insignificant a province in the mighty field of his creation? Are we the befitting objects of so great and so signal an interposition? Does not the largeness of that field which astronomy lays open to the view of modern science, throw a suspicion over the truth of the gospel history? and how shall we reconcile the greatness of that wonderful movement which was made in heaven for the redemption of fallen man, with the comparative meanness and obscurity of our species?

This is a popular argument against Christianity,

not much dwelt upon in books, but, we believe, a good deal insinuated in conversation, and having no small influence on the amateurs of a superficial philosophy. At all events, it is right that every such argument should be met, and manfully confronted; nor do we know a more discreditable surrender of our religion, than to act as if she had any thing to fear from the ingenuity of her most accomplished adversaries. The author of the following treatise engages in his present undertaking, under the full impression that a something may be found with which to combat Infidelity in all its forms; that the truth of God and of his message admits of a noble and decisive manifestation, through every mist which the pride, or the prejudice, or the sophistry of man may throw around it; and elevated as the wisdom of him may be, who has ascended the heights of science, and poured the light of demonstration over the most wondrous of nature's mysteries, that even out of his own principles it may be proved, how much more elevated is the wisdom of him who sits with the docility of a little child to his Bible, and casts down to its authority all his lofty imaginations.

DISCOURSE II.

THE MODESTY OF TRUE SCIENCE.

"And if any man think that he knoweth any thing, he knoweth nothing yet as he ought to know."—1 CORINTHIANS, viii. 2.

THERE is much profound and important wisdom in that proverb of Solomon, where it is said, that "the heart knoweth its own bitterness." It forms part of a truth still more comprehensive, that every man knoweth his own peculiar feelings, and difficulties, and trials, far better than he can get any of his neighbours to perceive them. It is natural to us all, that we should desire to engross, to the uttermost, the sympathy of others with what is most painful to the sensibilities of our own bosom, and with what is most aggravating in the hardships of our own situation. But, labour as we may, we cannot, with every power of expression, make an adequate conveyance, as it were, of all our sensations, and of all our circumstances, into another's understanding. There is a something in the intimacy of a man's own experience, which he cannot make to pass entire into the heart and mind even of his most familiar companion,—and thus it is, that he is so often defeated in his attempts to obtain a full and a cordial possession of his sympathy. He is mortified, and he wonders at the obtuseness of the people around him—and that he cannot get them to enter into the justness of his complainings —nor to feel the point upon which turn the truth and the reason of his remonstrances—nor to give their interested attention to the case of his peculiarities and of his wrongs—nor to kindle, in generous resentment, along with him, when he starts the topic of his indignation. He does not reflect, all the while, that, with every human being he addresses, there is an inner man, which forms a theatre of passions, and of interests as busy, as crowded, and as fitted as his own to engross the anxious and the exercised feelings of a heart, which can alone understand its own bitterness, and lay a correct estimate on the burden of its own visitations. Every man we meet, carries about with him, in the unperceived solitude of his bosom, a little world of his own—and we are just as blind, and as insensible, and as dull, both of perception and of sympathy, about his engrossing objects, as he is about ours; and, did we suffer this observation to have all its weight upon us, it might serve to make us more candid, and more considerate of others. It might serve to abate the monopolizing selfishness of our nature. It might serve to soften down all the malignity which comes out of those envious contemplations that we are so apt to cast on the fancied ease and prosperity which are around us. It might serve to reconcile every man to his own lot, and dispose him to bear, with thankfulness, his own burden; and if this train of sentiment were prosecuted with firmness, and calmness, and impartiality, it would lead to the conclusion, that each profession in life has its own peculiar pains, and its own besetting inconveniences—that, from the very bottom of society, up to the golden pinnacle which blazons upon its summit, there is much in the shape of care and of suffering to be found—that, throughout all the conceivable varieties of human condition, there are trials, which can neither be adequately told on the one side, nor fully understood on the other—that the ways of God to man are as equal in this, as in every department of his administration—and that, go to whatever quarter of human experience we may, we shall find that he has provided enough to exercise the patience, and to accomplish the purposes of a wise and a salutary discipline upon all his children.

I have brought forward this observation, that it may prepare the way for a second. There are perhaps no two sets of human beings, who comprehend less the movements, and enter less into the cares and concerns, of each other, than the wide and busy public on the one hand, and, on the other, those men of close and studious retirement, whom the world never hears of, save when, from their thoughtful solitude, there issues forth some splendid discovery, to set the world on a gaze of admiration. Then will the brilliancy of a superior genius draw every eye towards it—and the homage paid to intellectual superiority, will place its idol on a loftier eminence than all wealth or than all titles can bestow—and the name of the successful philosopher will circulate, in his own age, over the whole extent of civilized society, and be borne down to posterity in the characters of ever-during remembrance: and thus it is, that, when we look back on the days of

Newton, we annex a kind of mysterious greatness to him, who, by the pure force of his understanding, rose to such a gigantic elevation above the level of ordinary men—and the kings and warriors of other days sink into insignificance around him—and he, at this moment, stands forth to the public eye, in a prouder array of glory than circles the memory of all the men of former generations—and, while all the vulgar grandeur of other days is now mouldering in forgetfulness, the achievements of our great astronomer are still fresh in the veneration of his countrymen, and they carry him forward on the stream of time, with a reputation ever gathering, and the triumphs of a distinction that will never die.

Now, the point that I want to impress upon you is, that the same public, who are so dazzled and overborne by the lustre of all this superiority, are utterly in the dark as to what that is which confers its chief merit on the philosophy of Newton. They see the result of his labours, but they know not how to appreciate the difficulty or the extent of them. They look on the stately edifice he has reared, but they know not what he had to do in settling the foundation which gives to it all its stability; nor are they aware what painful encounters he had to make, both with the natural predilections of his own heart, and with the prejudices of others, when employed on the work of laying together its unperishing materials. They have never heard of the controversies which this man, of peaceful unambitious modesty, had to sustain with all that was proud, and all that was intolerant in the philosophy of the age. They have never, in thought, entered that closet which was the scene of his patient and profound exercises—nor have they gone along with him, as he gave his silent hours to the labours of the midnight oil, and plied that unwearied task, to which the charm of lofty contemplation had allured him—nor have they accompanied him through all the workings of that wonderful mind, from which, as from the recesses of a laboratory, there came forth such gleams and processes of thought as shed an effulgency over the whole amplitude of nature. All this, the public have not done; for of this the great majority, even of the reading and cultivated public, are utterly incapable; and therefore is it, that they need to be told what that is, in which the main distinction of his philosophy lies; that, when labouring in other fields of investigation, they may know how to borrow from his safe example, and how to profit by that superior wisdom which marked the whole conduct of his understanding.

Let it be understood, then, that they are the positive discoveries of Newton, which, in the eye of a superficial public, confer upon him all his reputation. He discovered the mechanism of the planetary system. He discovered the composition of light. He discovered the cause of those alternate movements which take place on the waters of the ocean. These form his actual and his visible achievements. These are what the world look to as the monuments of his greatness. These are doctrines by which he has enriched the field of philosophy; and thus it is, that the whole of his merit is supposed to lie in having had the sagacity to perceive, and the vigour to lay hold of the proofs, which conferred upon these doctrines all the establishment of a most rigid and conclusive demonstration.

But, while he gets all his credit, and all his admiration for those articles of science which he has added to the creed of philosophers, he deserves as much credit and admiration for those articles which he kept out of this creed, as for those which he introduced into it. It was the property of his mind, that it kept a tenacious hold of every one position which had proof to substantiate it: but it forms a property equally characteristic, and which, in fact, gives its leading peculiarity to the whole spirit and style of his investigations, that he put a most determined exclusion on every one position that was destitute of such proof. He would not admit the astronomical theories of those who went before him, because they had no proof. He would not give in to their notions about the planets wheeling their rounds in whirlpools of ether—for he did not see this ether—he had no proof of its existence: and, besides, even supposing it to exist, it would not have impressed, on the heavenly bodies, such movements as met his observation. He would not submit his judgment to the reigning systems of the day—for, though they had authority to recommend them, they had no proof; and thus it is, that he evinced the strength and the soundness of his philosophy, as much by his decisions upon those doctrines of science which he rejected, as by his demonstration of those doctrines of science which he was the first to propose, and which now stand out to the eye of posterity as the only monuments to the force and superiority of his understanding.

He wanted no other recommendation for any one article of science, than the recommendation of evidence—and, with this recommendation, he opened to it the chamber of his mind, though authority scowled upon it, and taste was disgusted by it, and fashion was ashamed of it, and all the beauteous speculation of former days was cruelly broken up by this new announcement of the better philosophy, and scattered like the fragments of an aërial vision, over which the past generations of the world had been slumbering their profound and their pleasing reverie. But, on the other hand, should the article of science want the recommendation of evidence, he shut against it all the avenues of his understanding—and though all antiquity lent their suffrages to it, and all eloquence had thrown around it the most attractive brilliancy, and all habit had incorporated it with every system of every seminary in Europe, and all fancy had arrayed it in graces of the most tempting solicitation; yet was the steady and inflexible mind of Newton proof against this whole weight of authority and allurement, and, casting his cold and unwelcome look at the specious plausibility, he rebuked it from his presence. The strength of his philosophy lay as much in refusing admittance to that which wanted evidence, as in giving a place and an occupancy to that which possessed it. In that march of intellect, which led him onwards through the rich and magnificent field of his discoveries, he pondered every step; and, while he advanced with a firm and assured movement, wherever the light of evidence carried him, he never suffered any glare of imagination or of prejudice to seduce him from his path.

Certain it is, that, in the prosecution of his wonderful career, he found himself on a way beset with temptation upon every side of him. It was not merely that he had the reigning taste and philosophy of the times to contend with. But he expatiated on a lofty region, where, in all the giddiness of success, he might have met with much to solicit his fancy, and tempt him to some devious speculation. Had he been like the majority of other men, he would have broken free from the fetters of a sober and chastised understanding, and, giving wing to his imagination, had done what philosophers have done after him—been carried away by some meteor of their own forming, or found their amusement in some of their own intellectual pictures, or palmed some loose and confident plausibilities of their own upon the world. But Newton stood true to his principle, that he would take up with nothing which wanted evidence, and he kept by his demonstrations, and his measurements, and his proofs; and, if it be true that he who ruleth his own spirit is greater than he who taketh a city, there was won, in the solitude of his chamber, many a repeated victory over himself, which should give a brighter lustre to his name than all the conquests he has made on the field of discovery, or than all the splendour of his positive achievements.

I trust you understand, that, though it be one of the maxims of the true philosophy, never to shrink from a doctrine which has evidence on its side, it is another maxim, equally essential to it, never to harbour any doctrine when this evidence is wanting. Take these two maxims along with you, and you will be at no loss to explain the peculiarity, which, more than any other, goes both to characterize and to ennoble the philosophy of Newton. What I allude to is, the precious combination of its strength and of its modesty. On the one hand, what greater evidence of strength than the fulfilment of that mighty enterprise, by which the heavens have been made its own, and the mechanism of unnumbered worlds has been brought within the grasp of the human understanding? Now, it was by walking in the light of sound and competent evidence, that all this was accomplished. It was by the patient, the strenuous, the unfaltering application of the legitimate instruments of discovery. It was by touching that which was tangible, and looking to that which was visible, and computing that which was measurable, and, in one word, by making a right and a reasonable use of all that proof which the field of nature around us has brought within the limit of sensible observation. This is the arena on which the modern philosophy has won all her victories, and fulfilled all her wondrous achievements, and reared all her proud and enduring monuments, and gathered all her magnificent trophies, to that power of intellect with which the hand of a bounteous heaven has so richly gifted the constitution of our species.

But, on the other hand, go beyond the limits of sensible observation, and, from that moment, the genuine disciples of this enlightened school cast all their confidence and all their intrepidity away from them. Keep them on the firm ground of experiment, and none more bold and more decisive in their announcements of all that they have evidence for—but, off this ground, none more humble, or more cautious of any thing like positive announcements, than they. They choose neither to know, nor to believe, nor to assert, where evidence is wanting, and they will sit, with all the patience of a scholar to his task, till they have found it. They are utter strangers to that haughty confidence with which some philosophers of the day sport the plausibilities of unauthorized speculation, and by which, unmindful of the limit that separates the region of sense from the region of conjecture, they make their blind and their impetuous inroads into a province which does not belong to them. There is no one object to which the exercised mind of a true Newtonian disciple is more familiarized than this limit, and it serves as a boundary by which he shapes, and bounds, and regulates all the enterprises of his philosophy. All the space which lies within this limit, he cultivates to the uttermost; and it is by such successive labours, that every year which rolls over the world is witnessing some new contribution to experimental science, and adding to the solidity and aggrandizement of this wonderful fabric. But, if true to their own principle, then, in reference to the forbidden ground which lies without this limit, those very men, who, on the field of warranted exertion, evinced all the hardihood and vigour of a full-grown understanding, show, on every subject where the light of evidence is withheld from them, all the modesty of children. They give us positive opinion only when they have indisputable proof—but, when they have no such proof, then they have no such opinion. The single principle of their respect to truth, secures their homage for every one position where the evidence of truth is present, and, at the same time, begets an entire diffidence about every one position from which this evidence is disjoined. And thus we may understand, how the first man in the accomplishments of philosophy, which the world ever saw, sat at the book of nature in the humble attitude of its interpreter and its pupil—how all the docility of conscious ignorance threw a sweet and softening lustre around the radiance even of his most splendid discoveries: and, while the flippancy of a few superficial acquirements is enough to place a philosopher of the day on the pedestal of his fancied elevation, and to vest him with an assumed lordship over the whole domain of natural and revealed knowledge; we cannot forbear to do honour to the unpretending greatness of Newton, than whom we know not if there ever lighted on the face of our world, one in the character of whose admirable genius so much force and so much humility were more attractively blended.

I now propose to carry you forward, by a few simple illustrations, to the argument of this day. All the sublime truths of the modern astronomy lie within the field of actual observation, and have the firm evidence to rest upon of all that information which is conveyed to us by the avenue of the senses. Sir Isaac Newton never went beyond this field, without a reverential impression upon his mind, of the precariousness of the ground on which he was standing. On this ground he never ventured a positive affirmation—but, resigning the lofty tone of demonstration, and putting on the modesty of conscious ignorance, he brought forward all he had to say in the humble form of a doubt, or a conjecture, or a question. But what he had not confidence to do, other philosophers have done after him—and they have winged their audacious way into forbidden regions—and they have crossed that circle by which the field of observation is enclosed—and there have they debated and dogmatized with all the pride of a most intolerant assurance.

Now, though the case be imaginary, let us conceive, for the sake of illustration, that one of these philosophers made so extravagant a departure from the sobriety of experimental science, as to pass on from the astronomy of the different planets, and to attempt the natural history of their animal and vegetable kingdoms. He might get hold of some vague and general analogies, to throw an air of plausibility around his speculation. He might pass from the botany of the different regions of the globe that we inhabit, and make his loose and confident applications to each of the other planets, according to its distance from the sun, and the inclination of its axis to the plane of its annual revolution; and out of some such slender materials, he may work up an amusing philosophical romance, full of ingenuity, and having, withal, the colour of truth and of consistency spread over it.

I can conceive how a superficial public might be

delighted by the eloquence of such a composition, and even be impressed by its arguments; but were I asked, which is the man of all the ages and countries in the world, who would have the least respect for this treatise upon the plants which grow on the surface of Jupiter, I should be at no loss to answer the question. I should say, that it would be he who had computed the motions of Jupiter—that it would be he who had measured the bulk and the density of Jupiter—that it would be he who had estimated the periods of Jupiter—that it would be he whose observant eye and patiently calculating mind, had traced the satellites of Jupiter through all the rounds of their mazy circulation, and unravelled the intricacy of all their movements. He would see at once that the subject lay at a hopeless distance beyond the field of legitimate observation. It would be quite enough for him, that it was beyond the range of his telescope. On this ground, and on this ground only, would he reject it as one of the puniest imbecilities of childhood. As to any character of truth or of importance, it would have no more effect on such a mind as that of Newton, than any illusion of poetry; and from the eminence of his intellectual throne, would he cast a penetrating glance at the whole speculation, and bid its gaudy insignificance away from him.

But let us pass onward to another case, which, though as imaginary as the former, may still serve the purpose of illustration.

This same adventurous philosopher may be conceived to shift his speculation from the plants of another world, to the character of its inhabitants.

He may avail himself of some slender correspondencies between the heat of the sun and the moral temperament of the people it shines upon. He may work up a theory, which carries on the front of it some of the characters of plausibility; but surely it does not require the philosophy of Newton to demonstrate the folly of such an enterprise. There is not a man of plain understanding, who does not perceive that this ambitious inquirer has got without his reach—that he has stepped beyond the field of experience, and is now expatiating on the field of imagination—that he has ventured on a dark unknown, where the wisest of all philosophy is the philosophy of silence, and a profession of ignorance is the best evidence of a solid understanding—that if he think he knows any thing on such a subject as this, "he knoweth nothing yet as he ought to know." He knows not what Newton knew, and what he kept a steady eye upon throughout the whole march of his sublime investigations. He knows not the limit of his own faculties. He has overleaped the barrier which hems in all the possibilities of human attainment. He has wantonly flung himself off from the safe and firm field of observation, and got on that undiscoverable ground, where, by every step he takes, he widens his distance from the true philosophy, and by every affirmation he utters, he rebels against the authority of all its maxims.

I can conceive it to be your feeling, that I have hitherto indulged in a vain expense of argument, and it is most natural for you to put the question, 'What is the precise point of convergence to which

I am directing all the light of this abundant and seemingly superfluous illustration?'

In the astronomical objection which Infidelity has proposed against the truth of the Christian revelation, there is first an assertion, and then an argument. The assertion is, that Christianity is set up for the exclusive benefit of our minute and solitary world. The argument is, that God would not lavish such a quantity of attention on so insignificant a field. Even though the assertion were admitted, I should have a quarrel with the argument. But the futility of the objection is not laid open in all its extent, unless we expose the utter want of all essential evidence even for the truth of the assertion. How do infidels know that Christianity is set up for the single benefit of this earth and its inhabitants? How are they able to tell us, that if you go to other planets, the person and the religion of Jesus are there unknown to them? We challenge them to the proof of this announcement. We see in this objection the same rash and gratuitous procedure, which was so apparent in the two cases that we have already advanced for the purpose of illustration. We see in it the same glaring transgression on the spirit and the maxims of that very philosophy which they profess to idolize. They have made their argument against us, out of an assertion which has positively no ascertained fact to rest upon—an assertion which they have no means whatever of verifying—an assertion, the truth or the falsehood of which can only be gathered out of some supernatural message, for it lies completely beyond the range of human observation. It is willingly admitted, that by an attempt at the botany of other worlds, the true method of philosophizing is trampled on; for this is a subject that lies beyond the range of actual observation, and every performance upon it must be made up of assertions without proofs. It is also willingly admitted, that an attempt at the civil and political history of their people, would be an equally extravagant departure from the spirit of the true philosophy; for this also lies beyond the field of actual observation; and all that could possibly be mustered up on such a subject as this, would still be assertions without proofs. Now, the theology of these planets is, in every way, as inaccessible a subject as their politics or their natural history; and therefore it is, that the objection, grounded on the confident assumption of those infidel astronomers, who assert Christianity to be the religion of this one world, or that the religion of these other worlds is not our very Christianity, can have no influence on a mind that has derived its habits of thinking, from the pure and rigorous school of Newton; for the whole of this assertion is just as glaringly destitute of proof, as in the two former instances.

The man who could embark in an enterprise so foolish and so fanciful, as to theorize on the details of the botany of another world, or to theorize on the natural and moral history of its people, is just making as outrageous a departure from all sense, and all science, and all sobriety, when he presumes to speculate, or to assert on the details or the methods of God's administration among its rational and accountable inhabitants. He wings his fa to as hazardous a region, and vainly strives a penetrating vision through the mantle of as deep an obscurity. All the elements of such a speculation are hidden from him. For any thing he can tell, sin has found its way into these other worlds. For any thing he can tell, their people have banished themselves from communion with God. For any thing he can tell, many a visit has been made to each of them, on the subject of our common Christianity, by commissioned messengers from the throne of the Eternal. For any thing he can tell, the redemption proclaimed to us is not one solitary instance, or not the whole of that redemption which is by the Son of God—but only our part in a plan of mercy, equal in magnificence to all that astronomy has brought within the range of human contemplation. For any thing he can tell, the moral pestilence, which walks abroad over the face of our world, may have spread its desolations over all the planets of all the systems which the telescope has made known to us. For any thing he can tell, some mighty redemption has been devised in heaven, to meet this disaster in the whole extent and malignity of its visitations. For any thing he can tell, the wonder-working God, who has strewed the field of immensity with so many worlds, and spread the shelter of His omnipotence over them, may have sent a message of love to each, and re-assured the hearts of its despairing people by some overpowering manifestation of tenderness. For any thing he can tell, angels from paradise may have sped to every planet their delegated way, and sung, from each azure canopy, a joyful annunciation, and said,

"Peace be to this residence, and good-will to all its families, and glory to Him in the highest, who, from the eminency of his throne, has issued an act of grace so magnificent, as to carry the tidings of life and of acceptance to the unnumbered orbs of a sinful creation." For any thing he can tell, the Eternal Son, of whom it is said, that by Him the worlds were created, may have had the government of many sinful worlds laid upon His shoulders; and by the power of His mysterious word, have awoke them all from that spiritual death, to which they had sunk in lethargy as profound as the slumbers of non-existence. For any thing he can tell, the one Spirit who moved on the face of the waters, and whose presiding influence it was that hushed the wild war of nature's elements, and made a beauteous system emerge out of its disjointed materials, may now be working with the fragments of another chaos; and educing order, and obedience, and harmony, out of the wrecks of a moral rebellion, which reaches through all these spheres, and spreads disorder to the uttermost limits of our astronomy.

But here I stop—nor shall I attempt to grope further my dark and fatiguing way, among such sublime and mysterious secrecies. It is not I who am offering to lift this curtain. It is not I who am pitching my adventurous flight to the secret things which belong to God, away from the things that are revealed, and which belong to us, and to our children. It is the champion of that very Infidelity which I am now combating. It is he who props his unchristian argument, by presumptions fetched out of those untravelled obscurities which lie on the other side of a barrier that I pronounce to be impassable. It is he who trangresses the limits which Newton forbore to enter; because, with a justness which reigns throughout all his inquiries, he saw the limit of his own understanding, nor would he venture himself beyond it. It is he who has borrowed from the philosophy of this wondrous man a few dazzling conceptions, which have only served to bewilder him—while, an utter stranger to the spirit of this philosophy, he has carried a daring and an ignorant speculation far beyond the boundary of its prescribed and allowable enterprises. It is he who has mustered against the truths of the Gospel, resting as it does on evidence within the reach of his faculties, an objection, for the truth of which he has no evidence whatever. It is he who puts away from him a doctrine, for which he has the substantial and the familiar proof of human testimony; and substitutes in its place, a doctrine, for which he can get no other support than from a reverie of his own imagination. It is he who turns aside from all that safe and certain argument, that is supplied by the history of this world, of which he knows something; and who loses himself in the work of theorizing about other worlds, of the moral and theological history of which he positively knows nothing. Upon him, and not upon us, lies the folly of launching his impetuous way beyond the province of observation—of letting his fancy afloat among the unknown of distant and mysterious regions—and, by an act of daring, as impious as it is unphilosophical, of trying to unwrap that shroud, which, till drawn aside by the hand of a messenger from heaven, will ever veil, from human eye, the purposes of the Eternal.

If you have gone along with us in the preceding observations, you will perceive how they are calculated to disarm of all its point, and of all its energy, that flippancy of Voltaire; when, in the examples he gives of the dotage of the human understanding, he tells us of Bacon having believed in witchcraft, and Sir Isaac Newton having written a commentary on the Book of Revelation. The former instance we shall not undertake to vindicate; but, in the latter instance, we perceive what this brilliant and specious, but withal superficial apostle of Infidelity, either did not see, or refused to acknowledge. We see in this intellectual labour of our great philosopher, the working of the very same principles which carried him through the profoundest and the most successful of his investigations; and how he kept most sacredly and most consistently by those very maxims, the authority of which, he, even in the full vigour and manhood of his faculties, ever recognized. We see in the theology of Newton, the very spirit and principle which gave all its stability, and all its sureness, to the philosophy of Newton. We see the same tenacious adherence to every one doctrine, that had such valid proof to uphold it, as could be gathered from the field of human experience; and we see the same firm resistance of every one argument, that had nothing to recommend it, but such plausibilities as could easily be devised by the genius of man, when he expatiated abroad on those fields of creation which the eye never witnessed, and from which no messenger ever came to us with any credible information. Now, it was on the former of these two principles that Newton clung so determinedly to his Bible, as the record of an actual annunciation from God to the inhabitants of this world. When he turned his attention to this book, he came to it with a mind tutored to the philosophy of facts—and when he looked at its credentials, he saw the stamp and the impress of this philosophy on every one of them. He saw the fact of Christ being a messenger from heaven, in the audible language by which it was conveyed from heaven's canopy to human ears. He saw the fact of his being an approved ambassador of God, in those miracles which carried their own resistless evidence along with them to human eyes. He saw the truth of this whole history brought home to his own conviction, by a sound and substantial vehicle of human testimony. He saw the reality of that supernatural light, which inspired the prophecies he himself illustrated, by such an agreement with the events of a various and distant futurity as could be taken cognizance of by human observation. He saw the wisdom of God pervading the whole substance of the written message, in such manifold adaptations to the circumstances of man, and to the whole secrecy of his thoughts, and his affections, and his spiritual wants, and his moral sensibilities, as even in the mind of an ordinary and unlettered peasant, can be attested by human consciousness. These formed the solid materials of the basis on which our experimental philosopher stood; and there was nothing in the whole compass of his own astronomy, to dazzle him away from it; and he was too well aware of the limit between what he knew, and what he did not know, to be seduced from the ground he had taken, by any of those brilliancies, which have since led so many of his humbler successors into the track of Infidelity. He had measured the distances of these planets. He had calculated their periods. He had estimated their figures, and their bulk, and their densities, and he had subordinated the whole intricacy of their movements to the simple and sublime agency of one commanding principle. But he had too much of the ballast of a substantial understanding about him, to be thrown afloat by all this success among the plausibilities of wanton and unauthorized speculation. He knew the boundary which hemmed him. He knew that he had not thrown one particle of light on the moral or religious history of these planetary regions. He had not ascertained what visits of communication they received from the God who upholds them. But he knew that the fact of a real visit made to this planet, had such evidence to rest upon, that it was not to be dispostеd by any aerial imagination. And when I look at the steady and unmoved Christianity of this wonderful man; so far from seeing any symptom of dotage and imbecility, or any forgetfulness of those principles on which the fabric of his philosophy is reared; do I see, that in sitting down to the work of a Bible commentator, he hath given as their most beautiful and most consistent exemplification.

I did not anticipate such a length of time, and of illustration, in this stage of my argument. But I will not regret it, if I have familiarized the minds of any of my readers to the reigning principle of this Discourse. We are strongly disposed to think, that it is a principle which might be made to apply to every argument of every unbeliever—and so to serve not merely as an antidote against the Infidelity of astronomers, but to serve as an antidote against all Infidelity. We are all aware of the diversity of complexion which Infidelity puts on. It looks one thing in the man of science and of liberal accomplishment. It looks another thing in the refined voluptuary. It looks still another thing in the common-place railer against the artifices of priestly domination. It looks another thing in the dark and unsettled spirit of him, whose every reflection is tinctured with gall, and who casts his envious and malignant scowl at all that stands associated with the established order of society. It looks another thing in the prosperous man of business, who has neither time nor patience for the details of the Christian evidence—but who, amid the hurry of his other occupations, has gathered as many of the lighter petulancies of the infidel writers, and caught from the perusal of them, as contemptuous a tone towards the religion of the New Testament, as to set him at large from all the decencies of religious observation, and to give him the disdain of an elevated complacency over all the follies of what he counts a vulgar superstition. And, lastly, for Infidelity has now got down amongst us to the humblest walks of life; may it occasionally be seen louring on the forehead of the resolute and hardy artificer, who can lift his menacing voice against the priesthood, and, looking on the Bible as a jugglery of theirs, can bid stout defiance to all its denunciations. Now, under all these varieties, we think that there might be detected the one and universal principle which we have attempted to expose. The something, whatever it is, which has dispossessed all these people of their Christianity, exists in their minds, in the shape of a position, which they hold to be true, but which, by no legitimate evidence, they have ever realized—and a position, which lodges within them as a wilful fancy or presumption of their own, but which could not stand the touchstone of that wise and solid principle, in virtue of which the followers of Newton give to observation the precedence over theory. It is a principle altogether worthy of being laboured—as, if carried round in faithful and consistent application amongst these numerous varieties, it is able to break up all the existing Infidelity of the world.

But there is one other most important conclusion to which it carries us. It carries us, with all the docility of children, to the Bible; and puts us down into the attitude of an unreserved surrender of thought and understanding, to its authoritative information. Without the testimony of an authentic messenger from Heaven, I know nothing of Heaven's counsels. I never heard of any moral telescope that can bring to my observation the doings or the deliberations which are taking place in the sanctuary of the Eternal. I may put into the registers of my belief, all that comes home to me through the senses of the outer man, or by the consciousness of the inner man. But neither the one nor the other can tell me of the purposes of God; can tell me of the transactions or the designs of His sublime monarchy; can tell me of the goings forth of Him who is from everlasting unto everlasting; can tell me of the march and the movements of that great administration which embraces all worlds, and takes into its wide and comprehensive survey the mighty roll of innumerable ages. It is true that my fancy may break its impetuous way into this lofty and inaccessible field; and, through the devices of my heart, which are many, the visions of an ever-shifting theology may take their alternate sway over me; but the counsel of the Lord, it shall stand. And I repeat it, that if true to the leading principle of that philosophy, which has poured such a flood of light over the mysteries of nature, we shall dismiss every self-formed conception of our own, and wait, in all the humility of conscious ignorance, till the Lord himself shall break His silence, and make His counsel known, by an act of communication. And now, that a professed communication is before me, and that it has all the solidity of the experimental evidence on its side, and nothing but the reveries of a daring speculation to oppose it, what is the consistent, what is the rational, what is the philosophical use that should be made of this document, but to set me down like a schoolboy, to the work of turning its pages, and conning its lessons, and submitting the every exercise of my judgment to its information and its testimony? We know that there is a superficial philosophy, which casts the glare of a most seducing brilliancy around it; and spurns the Bible, with all the doctrine, and all the piety of the Bible, away from it; and has infused the spirit of Antichrist into many of the literary establishments of the age; but it is not the solid, the profound, the cautious spirit of that philosophy, which has done so much to ennoble the modern period of our world; for the more that this spirit is cultivated and understood, the more will it be found in alliance with that spirit, in virtue of which all that exalteth itself against the knowledge of God is humbled, and all lofty imaginations are cast down, and every thought of the heart is brought into the captivity of the obedience of Christ.

DISCOURSE III.

ON THE EXTENT OF THE DIVINE CONDESCENSION.

"Who is like unto the Lord our God, who dwelleth on high; who humbleth himself to behold the things that are in heaven, and in the earth!"—PSALM cxiii. 5, 6.

In our last Discourse, we attempted to expose the total want of evidence for the assertion of the infidel astronomer—and this reduces the whole of our remaining controversy with him, to the business of arguing against a mere possibility. Still, however, the answer is not so complete as it might be, till the soundness of the argument be attended to, as well as the credibility of the assertion—or, in other words, let us admit the assertion, and take a view of the reasoning which has been constructed upon it.

We have already attempted to lay before you the wonderful extent of that space, teeming with unnumbered worlds, which modern science has brought within the circle of its discoveries. We even ventured to expatiate on those tracts of infinity, which lie on the other side of all that eye or that telescope hath made known to us—to shoot afar into those ulterior regions, which are beyond the limits of our astronomy—to impress you with the rashness of the imagination, that the creative energy of God had sunk exhausted by the magnitude of its efforts, at that very line, through which the art of man, lavished as it has been on the work of perfecting the instruments of vision, has not yet been able to penetrate; and upon all this we hazarded the assertion, that though all these visible heavens were to rush into annihilation, and the besom of the Almighty's wrath were to sweep from the face of the universe, those millions, and millions more of suns and of systems, which lie within the grasp of our actual observation—that this event, which, to our eye, would leave so wide and so dismal a solitude behind it, might be nothing in the eye of Him who could take in the whole, but the disappearance of a little speck from that field of created things, which the hand of His omnipotence had thrown around him.

But to press home the sentiment of the text, it is not necessary to stretch the imagination beyond the limit of our actual discoveries. It is enough to strike our minds with the insignificance of this world, and of all who inhabit it, to bring it into measurement with that mighty assemblage of worlds which lie open to the eye of man, aided as it has been by the inventions of his genius. When we told you of the eighty millions of suns, each occupying his own independent territory in space, and dispensing his own influences over a cluster of tributary worlds; this world could not fail to sink into littleness in the eye of him, who looked to all the magnitude and variety which are around it. We gave you but a feeble image of our comparative insignificance, when we said, that the glories of an extended forest would suffer no more from the fall of a single leaf, than the glories of this extended universe would suffer, though the globe we tread upon, " and all that it inherit, should dissolve." And when we lift our conceptions to Him who has peopled immensity with all these wonders—Who sits enthroned on the magnificence of His own works, and by one sublime idea can embrace the whole extent of that boundless amplitude, which He has filled with the trophies of His divinity; we cannot but resign our whole heart to the Psalmist's exclamation of " What is man, that thou art mindful of him; or the son of man, that thou shouldst deign to visit him!"

Now, mark the use to which all this has been turned by the genius of Infidelity. Such an humble portion of the universe as ours, could never have been the object of such high and distinguishing attentions as Christianity has assigned to it. God would not have manifested Himself in the flesh for the salvation of so paltry a world. The monarch of a whole continent would never move from his capital; and lay aside the splendour of royalty; and subject himself for months, or for years, to perils, and poverty, and persecution; and take up his abode in some small islet of his dominions, which, though swallowed by an earthquake, could not be missed amid the glories of so wide an empire; and all this to regain the lost affections of a few families upon its surface. And neither would the eternal Son of God—He who is revealed to us as having made all worlds, and as holding an empire, amid the splendours of which, the globe that we inherit is shaded in insignificance; neither would He strip Himself of the glory He had with the Father before the world was, and light on this lower scene for the purpose imputed to Him in the New Testament. Impossible, that the concerns of this puny ball, which floats its little round among an infinity of larger worlds, should be of such mighty account in the plans of the Eternal, or should have given birth in heaven to so wonderful a movement, as the Son of God putting on the form of our degraded species, and sojourning amongst us, and sharing in all our infirmities, and crowning the whole scene of humiliation by the disgrace and the agonies of a cruel martyrdom.

This has been started as a difficulty in the way of the Christian Revelation; and it is the boast of many of our philosophical Infidels, that, by the light of modern discovery, the light of the New Testament is eclipsed and overborne; and the mischief is not confined to philosophers, for the argument has got into other hands, and the popular illustrations that are now given to the sublimest truths of science, have widely disseminated all the Deism that has been grafted upon it; and the high tone of a decided contempt for the Gospel is now associated with the flippancy of superficial accquirements; and, while the venerable Newton, whose genius threw open those mighty fields of contemplation, found a fit exercise for his powers in the interpretation of the Bible, there are thousands and tens of thousands, who, though walking in the light which he holds out to them, are seduced by a complacency which he never felt, and inflated by a pride which never entered into his pious and philosophical bosom, and whose only notice of the Bible is to depreciate, and to deride, and to disown it.

Before entering into what we conceive to be the right answer to this objection, let us previously observe, that it goes to strip the Deity of an attribute, which forms a wonderful addition to the glories of his incomprehensible character. It is indeed a mighty evidence of the strength of His arm, that so many millions of worlds are suspended on it; but it would surely make the high attribute of His power more illustrious, if, while it expatiated at large among the suns and the systems of astronomy, it could, at the very same instant, be impressing a movement and a direction on all the minuter wheels of that machinery which is working incessantly around us. It forms a noble demonstration of His wisdom, that He gives unremitting operation to those laws which uphold the stability of this great universe; but it would go to heighten that wisdom inconceivably, if, while equal to the magnificent task of maintaining the order and harmony of the spheres, it was lavishing its inexhaustible resources on the beauties, and varieties, and arrangements, of every one scene, however humble, of every one field, however narrow, of the creation He had formed. It is a cheering evidence of the delight He takes in communicating happiness, that the whole of immensity should be so strewed with the habitations of life and of intelligence; but it would surely bring home the evidence, with a nearer and a more affecting impression, to every bosom, did we know, that at the very time His benignant regard took in the mighty circle of created beings, there was not a single family overlooked by Him, and that every individual in every corner of his dominions, was as effectually seen to, as if the object of an exclusive and undivided care. It is our imperfection, that we cannot give our attention to more than one object, at one and the same instant of time; but surely it would elevate our every idea of the perfections of God, did we know, that while his comprehensive mind could grasp the whole amplitude of nature, to the very outermost of its boundaries, He had an attentive eye fastened on the very humblest of its objects, and pondered every thought of my heart, and noticed every footstep of my goings, and treasured up in His remembrance every turn and every movement of my history.

And, lastly, to apply this train of sentiment to the matter before us; let us suppose that one among the countless myriads of worlds, should be visited by a moral pestilence, which spread through all its people, and brought them under the doom of a law, whose sanctions were unrelenting and immutable; it were no disparagement to God, should He, by an act of righteous indignation, sweep this offence away from the universe which it deformed—nor should we wonder, though, among the multitude of other worlds, from which the ear of the Almighty was regaled with the songs of praise, and the incense of a pure adoration ascended to His throne, He should leave the strayed and solitary world to perish in the guilt of its rebellion. But, would it not throw the softening of a most exquisite tenderness over the character of God, should we see Him putting forth His every expedient to reclaim to Himself those children who had wandered away from Him—and, few as they were when compared with the host of His obedient worshippers, would it not just impart to his attribute of compassion the infinity of the Godhead, that, rather than lose the single world which had turned to its own way, He should send the messengers of peace to woo and to welcome it back again; and, if justice demanded so mighty a sacrifice, and the law behoved to be so magnified and made honourable, would it not throw a moral sublime over the goodness of the Deity, should He lay upon His own Son the burden of its atonement, that He might again smile upon the world, and hold out the sceptre of invitation to all its families?

We avow it, therefore, that this infidel argument goes to expunge a perfection from the character of God. The more we know of the extent of nature, should not we have the loftier conception of Him who sits in high authority over the concerns of so wide a universe? But is it not adding to the bright catalogue of His other attributes, to say, that, while magnitude does not overpower Him, minuteness cannot escape Him, and variety cannot bewilder Him; and that, at the very time while the mind of the Deity is abroad over the whole vastness of creation, there is not one particle of matter, there is not one individual principle of rational or of animal existence, there is not one single world in that expanse which teems with them, that His eye does not discern as constantly, and His hand does not guide as unerringly, and His Spirit does not watch and care for as vigilantly, as if it formed the one and exclusive object of His attention?

The thing is inconceivable to us, whose minds are so easily distracted by a number of objects, and this is the secret principle of the whole Infidelity I am now alluding to. To bring God to the level of our own comprehension, we would clothe him in the impotency of a man. We would transfer to his wonderful mind all the imperfection of our own faculties. When we are taught by astronomy, that He has millions of worlds to look after, and thus add in one direction to the glories of His character; we take away from them in another, by saying, that each of these worlds must be looked after imperfectly. The use that we make of a discovery, which should heighten our every conception of God, and humble us into the sentiment, that a Being of such mysterious elevation is to us unfathomable, is to sit in judgment over Him, and to pronounce such a judgment as degrades Him, and keeps Him down to the standard of our own paltry imagination! We are introduced by modern science to a multitude of other suns and of other systems; and the perverse interpretation we put upon the fact, that God *can* diffuse the benefits of His power and of His goodness over such a variety of worlds, is, that He *cannot*, or will not, bestow so much goodness on one of those worlds, as a professed revelation from Heaven has announced to us. While we enlarge the provinces of His empire, we tarnish all the glory of this enlargement, by saying, He has so much to care for, that the care of every one province must be less complete, and less vigilant, and less effectual, than it would otherwise have been. By the discoveries of modern science, we multiply the places of the creation; but along with this, we would impair the attribute of His eye being in every place to behold the evil and the good; and thus, while we magnify one of His perfections, we do it at the expense of another; and, to bring Him within the grasp of our feeble capacity, we would deface one of the glories of that character, which it is our part to adore, as higher than all thought, and as greater than all comprehension.

The objection we are discussing, I shall state again in a single sentence. Since astronomy has unfolded to us such a number of worlds, it is not likely that God would pay so much attention to this one world, and set up such wonderful provisions for its benefit, as are announced to us in the Christian Revelation. This objection will have received its answer, if we can meet it by the following position: —that God, in addition to the bare faculty of dwelling on a multiplicity of objects at one and the same time, has this faculty in such wonderful perfection, that He can attend as fully, and provide as richly, and manifest all His attributes as illustriously, on every one of these objects, as if the rest had no existence, and no place whatever in His government or in His thoughts.

For the evidence of this position, we appeal, in the first place, to the personal history of each individual among you. Only grant us, that God never loses sight of any one thing He has created, and that no created thing can continue either to be, or to act independently of Him; and then, even upon the face of this world, humble as it is on the great scale of astronomy, how widely diversified, and how multiplied into many thousand distinct exercises, is the attention of God! His eye is upon every hour of my existence. His spirit is intimately present with every thought of my heart. His inspiration gives birth to every purpose within me. His hand impresses a direction on every footstep of my goings. Every breath I inhale, is drawn by an energy which God deals out to me. This body, which, upon the slightest derangement, would become the prey of death, or of woful suffering, is now at ease, because He at this moment is warding off from me a thousand dangers, and upholding the thousand movements of its complex and delicate machinery. His presiding influence keeps by me through the whole current of my restless and everchanging history. When I walk by the wayside, He is along with me. When I enter into company, amid all my forgetfulness of Him, He never forgets me. In the silent watches of the night, when my eyelids have closed, and my spirit has sunk into unconsciousness, the observant eye of Him who never slumbers is upon me. I cannot fly from his presence. Go where I will, He tends me, and watches me, and cares for me; and the same Being who is now at work in the remotest domains of Nature and of Providence, is also at my right hand to eke out to me every moment of my being, and to uphold me in the exercise of all my feelings, and of all my faculties.

Now, what God is doing with me, He is doing with every distinct individual of this world's population. The intimacy of His presence, and attention, and care, reaches to one and to all of them. With a mind unburdened by the vastness of all its other concerns, He can prosecute, without distraction, the government and guardianship of every one son and daughter of the species. And is it for us, in the face of all this experience, ungratefully to draw a limit around the perfections of God —to aver, that the multitude of other worlds has withdrawn any portion of His benevolence from the one we occupy—or that He, whose eye is upon every separate family of the earth, would not lavish all the riches of His unsearchable attributes on some high plan of pardon and immortality, in behalf of its countless generations?

But, secondly, were the mind of God so fatigued, and so occupied with the care of other worlds, as the objection presumes Him to be, should we not see some traces of neglect, or of carelessness, in His management of ours? Should we not behold, in many a field of observation, the evidence of its master being over-crowded with the variety of His other engagements? A man oppressed by a multitude of business, would simplify and reduce the work of any new concern that was devolved upon him. Now, point out a single mark of God being thus oppressed. Astronomy has laid open to us so many realms of creation, which were before unheard of, that the world we inhabit shrinks into one remote and solitary province of His wide monarchy. Tell us then, if, in any one field of this province which man has access to, you witness a single indication of God sparing Himself—of

God reduced to languor by the weight of His other employments—of God sinking under the burden of that vast superintendance which lies upon Him—of God being exhausted, as one of ourselves would be, by any number of concerns, however great, by any variety of them, however manifold; and do you not perceive, in that mighty profusion of wisdom and of goodness, which is scattered every where around us, that the thoughts of this unsearchable Being are not as our thoughts, nor his ways as our ways?

My time does not suffer me to dwell on this topic, because, before I conclude, I must hasten to another illustration. But when I look abroad on the wondrous scene that is immediately before me—and see, that in every direction it is a scene of the most various and unwearied activity—and expatiate on all the beauties of that garniture by which it is adorned, and on all the prints of design and of benevolence which abound in it—and think, that the same God who holds the universe, with its every system, in the hollow of His hand, pencils every flower, and gives nourishment to every blade of grass, and actuates the movements of every living thing, and is not disabled, by the weight of His other cares, from enriching the humble department of nature I occupy, with charms and accommodations of the most unbounded variety—then, surely, if a message, bearing every mark of authenticity, should profess to come to me from God, and inform me of his mighty doings for the happiness of our species, it is not for me, in the face of all this evidence, to reject it as a tale of imposture, because astronomers have told me that He has so many other worlds and other orders of beings to attend to,—and, when I think that it were a deposition of Him from his supremacy over the creatures he has formed, should a single sparrow fall to the ground without His appointment, then let science and sophistry try to cheat me of my comfort, as they may—I will not let go the anchor of my confidence in God—I will not be afraid, for I am of more value than many sparrows.

But, thirdly, it was the telescope, that, by piercing the obscurity which lies between us and distant worlds, put Infidelity in possession of the argument against which we are now contending. But, about the time of its invention, another instrument was formed, which laid open a scene no less wonderful, and rewarded the inquisitive spirit of man with a discovery, which serves to neutralize the whole of this argument. This was the microscope. The one led me to see a system in every star. The other leads me to see a world in every atom. The one taught me, that this mighty globe, with the whole burden of its people and of its countries, is but a grain of sand on the high field of immensity. The other teaches me, that every grain of sand may harbour within it the tribes and the families of a busy population. The one told me of the insignificance of the world I tread upon. The other redeems it from all its insignificance; for it tells me that in the leaves of every forest, and in the flowers of every garden, and in the waters of every rivulet, there are worlds teeming with life, and numberless as are the glories of the firmament. The one has suggested to me, that beyond and above all that is visible to man, there may lie fields of creation which sweep immeasurably along, and carry the impress of the Almighty's hand to the remotest scenes of the universe. The other suggests to me, that within and beneath all that minuteness which the aided eye of man has been able to explore, there may lie a region of invisibles; and that, could we draw aside the mysterious curtain which shrouds it from our senses, we might there see a theatre of as many wonders as astronomy has unfolded, a universe within the compass of a point so small, as to elude all the powers of the microscope, but where the wonder-working God finds room for the exercise of all His attributes, where He can raise another mechanism of worlds, and fill and animate them all with the evidences of His glory.

Now, mark how all this may be made to meet the argument of our infidel astronomers. By the telescope, they have discovered that no magnitude, however vast, is beyond the grasp of the Divinity. But by the microscope, we have also discovered, that no minuteness, however shrunk from the notice of the human eye, is beneath the condescension of His regard. Every addition to the powers of the one instrument, extends the limit of His visible dominions. But, by every addition to the powers of the other instrument, we see each part of them more crowded than before, with the wonders of His unwearying hand. The one is constantly widening the circle of His territory. The other is as constantly filling up its separate portions, with all that is rich, and various, and exquisite. In a word, by the one I am told that the Almighty is now at work in regions more distant than geometry has ever measured, and among worlds more manifold than numbers have ever reached. But, by the other, I am also told, that with a mind to comprehend the whole, in the vast compass of its generality, He has also a mind to concentrate a close and a separate attention on each and on all of its particulars; and that the same God, who sends forth an upholding influence among the orbs and the movements of astronomy, can fill the recesses of every single atom with the intimacy of his presence, and travel, in all the greatness of His unimpaired attributes, upon every one spot and corner of the universe He has formed

They, therefore, who think that God will not put forth such a power, and such a goodness, and such a condescension, in behalf of this world, as are ascribed to Him in the New Testament, because He has so many other worlds to attend to, think of him as a man. They confine their view to the informations of the telescope, and forget altogether the informations of the other instrument. They only find room in their minds for His one attribute of a large and general superintendence; and keep out of their remembrance the equally impressive proofs we have for his other attribute, of a minute and multiplied attention to all that diversity of operations, where it is He that worketh all in all. And when I think, that as one of the instruments of philosophy has heightened our every impression of the first of these attributes, so another instrument has no less heightened our impression of the second of them—then I can no longer resist the conclusion, that it would be a transgression of sound argument, as well as a daring of impiety, to draw a limit around the doings of this unsearchable God—and, should a professed revelation from heaven tell me of an act of condescension, in behalf of some separate world, so wonderful, that angels desired to look into it, and the Eternal Son had to move from His seat of glory to carry it into accomplishment, all I ask is the evidence of such a revelation; for, let it tell me as much as it may of God letting himself down for the benefit of one single province of His dominions, this is no more than what I see lying scattered, in numberless examples, before me; and running through the whole line of my recollections; and meeting me in every walk of observation to which I can betake myself; and, now that the microscope has unveiled the wonders of another region, I see strewed around me, with a profusion which baffles my every attempt to comprehend it, the evidence that there is no one portion of the universe of God too minute for His notice, nor too humble for the visitations of His care.

As the end of all these illustrations, let me bestow a single paragraph on what I conceive to be the precise state of this argument.

It is a wonderful thing that God should be so unencumbered by the concerns of a whole universe, that He can give a constant attention to every moment of every individual in this world's population. But, wonderful as it is, you do not hesitate to admit it as true, on the evidence of your own recollections. It is a wonderful thing that He, whose eye is at every instant on so many worlds, should have peopled the world we inhabit with all the traces of the varied design and benevolence which abound in it. But great as the wonder is, you do not allow so much as the shadow of improbability to darken it, for its reality is what you actually witness, and you never think of questioning the evidence of observation. It is wonderful, it is passing wonderful, that the same God, whose presence is diffused through immensity, and who spreads the ample canopy of His administration over all its dwelling-places, should, with an energy as fresh and as unexpended as if He had only begun the work of creation, turn Him to the neighbourhood around us, and lavish, on its every hand-breadth, all the exuberance of His goodness, and crowd it with the many thousand varieties of conscious existence. But, be the wonder incomprehensible as it may, you do not suffer in your mind the burden of a single doubt to lie upon it, because you do not question the report of the microscope. You do not refuse its information, nor turn away from it as an incompetent channel of evidence. But to bring it still nearer to the point at issue, there are many who never looked through a microscope, but who rest an implicit faith in all its revelations; and upon what evidence I would ask? Upon the evidence of testimony—upon the credit they give to the authors of the books they have read, and the belief they put in the record of their observations. Now, at this point I make my stand. It is wonderful that God should be so interested in the redemption of a single world, as to send forth his well-beloved Son upon the errand; and He, to accomplish it, should, mighty to save, put forth all His strength, and travail in the greatness of it. But such wonders as these have already multiplied upon you; and when evidence is given of their truth, you have resigned your every judgment of the unsearchable God, and rested in the faith of them. I demand, in the name of sound and consistent philosophy, that you do the same in the matter before us—and take it up as a question of evidence—and examine that medium of testimony through which the miracles and informations of the Gospel have come to your door—and go not to admit as argument here, what would not be admitted as argument in any of the analogies of nature and observation—and take along with you in this field of inquiry, a lesson which you should have learned upon other fields—even the depth of the riches both of the wisdom and the knowledge of God, that His judgments are unsearchable, and His ways are past finding out.

I do not enter at all into the positive evidence for the truth of the Christian Revelation, my single aim at present being to dispose of one of the objections which is conceived to stand in the way of it. Let me suppose then, that this is done to the satisfaction of a philosophical inquirer; and that the evidence is sustained; and that the same mind that is familiarized to all the sublimities of natural science, and has been in the habit of contemplating God in association with all the magnificence which is around him, shall be brought to submit its thoughts to the captivity of the doctrine of Christ. Oh! with what veneration, and gratitude, and wonder, should he look on the descent of Him into this lower world, who made all these things, and without whom was not any thing made that was made. What a grandeur does it throw over every step in the redemption of a fallen world, to think of its being done by Him who unrobed Him of the glories of so wide a monarchy, and came to this humblest of its provinces, in the disguise of a servant, and took upon Him the form of our degraded species, and let Himself down to sorrows, and to sufferings, and to death, for us! In this love of an expiring Saviour to those for whom in agony He poured out His soul, there is a height, and a depth, and a length, and a breadth, more than I can comprehend; and let me never from this moment neglect so great a salvation, or lose my hold of an atonement, made sure by Him who cried that it was finished, and brought in an everlasting righteousness. It was not the visit of an empty parade that He made to us. It was for the accomplishment of some substantial purpose; and if that purpose is announced, and stated to consist in His dying the just for the unjust, that He might bring us unto God, let us never doubt of our acceptance in that way of communication with our Father in heaven, which he hath opened and made known to us. In taking to that way, let us follow His every direction, with that humility which a sense of all this wonderful condescension is fitted to inspire. Let us forsake all that He bids us forsake. Let us do all that He bids us do. Let us give ourselves up to his guidance with the docility of children overpowered by a kindness that we never merited, and a love that is unequalled by all the perverseness and all the ingratitude of our stubborn nature—for what shall we render unto Him for such mysterious benefits—to him who has thus been mindful of us—to him who thus has deigned to visit us?

But the whole of this argument is not yet exhausted. We have scarcely entered on the defence that is commonly made against the plea which Infidelity rests on the wonderful extent of the universe of God, and the insignificancy of our assigned portion of it. The way in which we have attempted to dispose of this plea, is by insisting on the evidence that is every where around us, of God combining, with the largeness of a vast and mighty superintendance, which reaches the outskirts of creation, and spreads over all its amplitudes—the faculty of bestowing as much attention, and exercising as complete and manifold a wisdom, and lavishing as profuse and inexhaustible a goodness, on each of its humblest departments, as if it formed the whole extent of His territory.

In the whole of this argument we have looked upon the earth as isolated from the rest of the universe altogether. But, according to the way in which the astronomical objection is commonly met, the earth is not viewed as in a state of detachment from the other worlds, and the other orders of being which God has called into existence. It is looked upon as the member of a more extended system. It is associated with the magnificence of a moral empire, as wide as the kingdom of nature. It is not merely asserted, what in our last Discourse has been already done, that for any thing we can know by reason, the plan of redemption may have its influences and its bearings on those creatures of God who people other regions, and occupy other fields in the immensity of his dominions; that to argue, therefore, on this plan being instituted for the single benefit of the world we live in, and of the species to which we belong, is a mere presumption of the Infidel himself; and that the objection he rears on it must fall to the ground, when the vanity of the presumption is exposed. The Christian apologist thinks he can go farther than this—that he can not merely expose the utter baselessness of the Infidel assertion, but that he has positive ground for erecting an opposite and a confronting assertion in its place—and that, after having neutralized their position, by showing the entire absence of all observation in its behalf, he can pass on to the distinct and affirmative testimony of the Bible.

We do think that this lays open a very interesting track, not of wild and fanciful, but of most legitimate and sober-minded speculation. And anxious as we are to put every thing that bears upon the Christian argument, into all its lights; and fearless as we feel for the result of a most thorough sifting of it; and thinking as we do think it, the foulest scorn that any pigmy philosopher of the day sho ld mince his ambiguous scepticism to a set of gi dy and ignorant admirers, or that a half-learned and superficial public should associate with the Christian priesthood, the blindness and the bigotry of a sinking cause—with these feelings we are not disposed to shun a single question that may be started on the subject of the Christian evidences. There is not one of its parts or bearings which needs the shelter of a disguise thrown over it. Let the priests of another faith ply their prudential expedients, and look so wise and so wary in the execution of them. But Christianity stands in a higher and a firmer attitude. The defensive armour of a shrinking or timid policy does not suit her. Hers is the naked majesty of truth; and with all the grandeur of age, but with none of its infirmities, has she come down to us, and gathered new strength from the battles she has won in the many controversies of many generations. With such a religion as this there is nothing to hide. All should be above boards. And the broadest light of day should be made fully and freely to circulate throughout all her secrecies. But secrets she has none. To her belong the frankness and the simplicity of conscious greatness; and whether she has to contend with the pride of philosophy, or stand in fronted opposition to the prejudices of the multitude, she does it upon her own strength, and spurns all the props and all the auxiliaries of superstition away from her.

DISCOURSE IV.

ON THE KNOWLEDGE OF MAN'S MORAL HISTORY IN THE DISTANT PLACES OF CREATION

"Which things the angels desire to look into."—1 PETER i. 12.

THERE is a limit, across which man cannot carry any one of his perceptions, and from the ulterior of which he cannot gather a single observation to guide or to inform him. While he keeps by the objects which are near, he can get the knowledge of them conveyed to his mind through the ministry of several of the senses. He can feel a substance that is within reach of his hand. He can smell a flower that is presented to him. He can taste the food that is before him. He can hear a sound of certain pitch and intensity; and, so much does this sense of hearing widen his intercourse with external nature, that, from the distance of miles, it can bring him in an occasional intimation.

But of all the tracts of conveyance which God has been pleased to open up between the mind of man, and the theatre by which he is surrounded, there is none by which he so multiplies his acquaintance with the rich and the varied creation on every side of him, as by the organ of the eye. It is this which gives to man his loftiest command over the scenery of nature. It is this by which so broad a range of observation is submitted to him. It is this which enables him, by the act of a single moment, to send an exploring look over the surface of an ample territory, to crowd his mind with the whole assembly of its objects, and to fill his vision with those countless hues which diversify and adorn it. It is this which carries him abroad over all that is sublime in the immensity of distance; which sets him as it were on an elevated platform, from whence he may cast a surveying glance over the arena of innumerable worlds; which spreads before him so mighty a province of contemplation, that the earth he inhabits only appears to furnish him with the pedèstal on which he may stand, and from which he may descry the wonders of all that magnificence which the Divinity has poured so abundantly around him. It is by the narrow outlet of the eye, that the mind of man takes its excursive flight over those golden tracks, where, in all the exhaustlessness of creative wealth, lie scattered the suns and the systems of astronomy. But how good a thing it is, and how becoming well, for the philosopher to be humble even amid the proudest march of human discovery, and the sublimest triumphs of the human understanding, when he thinks of that unscaled barrier, beyond which no power, either of eye or of telescope, shall ever carry him; when he thinks that, on the other side of it, there is a height, and a depth, and a length, and a breadth, to which the whole of this concave and visible firmament dwindles into the insignificancy of an atom—and above all, how ready should he be to cast every lofty imagination away from him, when he thinks of the God, who, on the simple foundation of His word, has reared the whole of this stately architecture, and, by the force of His preserving hand, continues to uphold it; and should the word again come out from Him, that this earth shall pass away, and a portion of the heavens which are around it, shall fall back into the annihilation from which He at first summoned them—what an impressive rebuke does it bring on the swelling vanity of science, to think that the whole field of its most ambitious enterprises may be swept away altogether, and still there remain before the eye of Him who sitteth on the throne, an untravelled immensity, which He hath filled with innumerable splendours, and over the whole face of which he hath inscribed the evidence of His high attributes, in all their might, and in all their manifestation.

But man has a great deal more to keep him humble of his understanding, than a mere sense of that boundary which skirts and which terminates the material field of his contemplations. He ought also to feel, how, within that boundary, the vast majority of things is mysterious and unknown to him—that even in the inner chamber of his own consciousness, where so much lies hidden from the observation of others, there is also to himself a little world of incomprehensibles; that if stepping beyond the limits of this familiar home, he look no farther than to the members of his family, there is much in the cast and the colour of every mind that is above his powers of divination; that in proportion as he recedes from the centre of his own personal experience, there is a cloud of ignorance and secrecy which spreads, and thickens, and throws a deep and impenetrable veil over the intricacies of every one department of human contemplation; that of all around him, his knowledge is naked and superficial, and confined to a few of those more conspicuous lineaments which strike upon his senses; that the whole face, both of nature and of society, presents him with questions which he cannot unriddle, and tells him that beneath the surface of all that the eye can rest upon, there lies the profoundness of a most unsearchable latency; and should he in some lofty enterprise of thought, leave this world, and shoot afar into those tracks of speculation which astronomy has opened, should he, baffled by the mysteries which beset his footsteps upon earth, attempt an ambitious flight towards the mysteries of heaven—let him go, but let the justness of a pious and philosophical modesty go along with him —let him forget not, that from the moment his mind has taken its ascending way for a few little miles above the world he treads upon, his every sense abandons him but one—that number, and motion, and magnitude, and figure, make up all the bareness of its elementary informations—that these orbs have sent him scarce another message than told by their feeble glimmering upon his eye, the simple fact of their existence—that he sees not the landscape of other worlds—that he knows not the moral system of any one of them—nor athwart the long and trackless vacancy which lies between, does there fall upon his listening ear the hum of their mighty populations.

But the knowledge which he cannot fetch up himself from the obscurity of this wondrous but untravelled scene, by the exercise of any one of his own senses, might be fetched to him by the testimony of a competent messenger. Conceive a native of one of these planetary mansions to light upon our world; and all we should require, would be, to be satisfied of his credentials, that we may give our faith to every point of information he had to offer us. With the solitary exception of what we have been enabled to gather by the instruments of astronomy, there is not one of his communications about the place he came from, on which we possess any means at all of confronting him; and, therefore, could he only appear before us invested with the characters of truth, we should never think of any thing else than taking up the whole matter of his testimony just as he brought it to us.

It were well had a sound philosophy schooled its professing disciples to the same kind of acquiescence in another message, which has actually come to the world; and has told us of matters still more remote from every power of unaided observation; and has been sent from a more sublime and mysterious distance, even from that God of whom it is said, that "clouds and darkness are the habitation of his throne;" and treating of a theme so lofty and so inaccessible, as the counsels of that Eternal Spirit, "whose goings forth are of old, even from everlasting," challenges of man that he should submit his every thought to the authority of this high communication. Oh! had the philosophers of the day known as well as their great master, how to draw the vigorous land-mark which verges the field of legitimate discovery, they should have seen when it is that philosophy becomes vain, and science is falsely so called; and how it is, that when philosophy is true to her principles, she shuts up her faithful votary to the Bible, and makes him willing to count all but loss, for the knowledge of Jesus Christ, and of Him crucified.

But let it be well observed, that the object of this message is not to convey information to us about the state of these planetary regions. This is not the matter with which it is fraught. It is a message from the throne of God to this rebellious province of His dominions; and the purpose of it is, to reveal the fearful extent of our guilt and of our danger, and to lay before us the overtures of reconciliation. Were a similar message sent from the metropolis of a mighty empire to one of its remote and revolutionary districts, we should not look to it for much information about the state or economy of the intermediate provinces. This were a departure from the topic on hand—though still there may chance to be some incidental allusions to the extent and resources of the whole monarchy, to the existence of a similar spirit of rebellion in other quarters of the land, or to the general principle of loyalty by which it was pervaded. Some casual references of this kind may be inserted in such a proclamation, or they may not—and it is with this precise feeling of ambiguity that we open the record of that embassy which has been sent us from heaven, to see if we can gather any thing there, about other places of the creation, to meet the objections of the infidel astronomer. But, while we pursue this object, let us be careful not to push the speculation beyond the limits of the written testimony; let us keep a just and a steady eye on the actual boundary of our knowledge, that, throughout every distinct step of our argument, we might preserve that chaste and unambitious spirit, which characterizes the philosophy of him who explored these distant heavens, and, by the force of his genius, unravelled the secret of that wondrous mechanism which upholds them.

The informations of the Bible upon this subject, are of two sorts—that from which we confidently gather the fact, that the history of the redemption of our species is known in other and distant places of the creation—and that from which we indistinctly guess at the fact, that the redemption itself may stretch beyond the limits of the world we occupy.

And here it may shortly be adverted to, that, though we know little or nothing of the moral and theological economy of the other planets, we are not to infer, that the beings who occupy these widely extended regions, even though not higher than we in the scale of understanding, know little of ours. Our first parents, ere they committed that act by which they brought themselves and their posterity into the need of redemption, had frequent and familiar intercourse with God. He walked with them in the garden of paradise, and there did angels hold their habitual converse; and, should the same unblotted innocence which charmed and attracted these superior beings to the haunts of Eden, be perpetuated in every planet but our own, then might each of them be the scene of high and heavenly communications, and an open way for the messengers of God be kept up with them all, and their inhabitants be admitted to a share in the themes and contemplations of angels, and have their spirits exercised on those things, of which we are told that the angels desired to look into them; and thus, as we talk of the public mind of a city, or the public mind of an empire—by the well-frequented avenues of a free and ready circulation, a public mind might be formed throughout the whole extent of God's sinless and intelligent creation—and, just as we often read of the eyes of all Europe being turned to the one spot where some affair of eventful importance is going on, there might be the eyes of a whole universe turned to the one world, where rebellion against the Majesty of heaven had planted its standard; and for the readmission of which within the circle of His fellowship, God, whose justice was inflexible, but whose mercy He had, by some plan of mysterious wisdom, made to rejoice over it, was putting forth all the might, and travailing in all the greatness of the attributes which belonged to Him.

But, for the full understanding of this argument, it must be remarked, that while in our exiled habitation, where all is darkness, and rebellion, and enmity, the creature engrosses every heart; and our affections, when they shift at all, only wander from one fleeting vanity to another, it is not so in the habitations of the unfallen. There, every desire and every movement is subordinated to God. He is seen in all that is formed, and in all that is spread around them—and, amid the fulness of that delight with which they expatiate over the good and the fair of this wondrous universe, the animating charm which pervades their every contemplation, is, that they behold, on each visible thing, the impress of the mind that conceived, and of the hand that made and that upholds it. Here, God is banished from the thoughts of every natural man, and, by a firm and constantly maintained act of usurpation, do the things of sense and of time wield an entire ascendancy. There, God is all in all. They walk in His light. They rejoice in the beatitudes of His presence. The veil is from off their eyes; and they see the character of a presiding Divinity in every scene, and in every event to which the Divinity has given birth. It is this which stamps a glory and an importance on the whole field of their contemplations; and when they see a new evolution in the history of created things, the reason they bend towards it so attentive an eye, is, that it speaks to their understanding some new evolution in the purposes of God—some new manifestation of His high attributes—some new and interesting step in the history of His sublime administration.

Now, we ought to be aware how it takes off, not from the intrinsic weight, but from the actual impression of our argument, that this devotedness to God which reigns in other places of the creation; this interest in Him as the constant and essential principle of all enjoyment; this concern in the untaintedness of his glory; this delight in the survey of His perfections and His doings, are what the men of our corrupt and darkened world cannot sympathize with.

But however little we may enter into it, the Bible

tells us, by many intimations, that amongst those creatures who have not fallen from their allegiance, nor departed from the living God, God is their all —that love to Him sits enthroned in their hearts, and fills them with all the ecstasy of an overwhelming affection—that a sense of grandeur never so elevates their souls, as when they look at the might and majesty of the Eternal—that no field of cloudless transparency so enchants them by the blissfulness of its visions, as when, at the shrine of infinite and unspotted holiness, they bend themselves in raptured adoration—that no beauty so fascinates and attracts them, as does that moral beauty which throws a softening lustre over the awfulness of the Godhead —in a word, that the image of his character is ever present to their contemplations, and the unceasing joy of their sinless existence lies in the knowledge and the admiration of Deity.

Let us put forth an effort, and keep a steady hold of this consideration, for the deadness of our earthly imaginations makes an effort necessary; and we shall perceive, that though the world we live in were the alone theatre of redemption, there is a something in the redemption itself that is fitted to draw the eye of an arrested universe towards it. Surely, where delight in God is the constant enjoyment, and the earnest intelligent contemplation of God is the constant exercise, there is nothing in the whole compass of nature or of history, that can so set His adoring myriads upon the gaze, as some new and wondrous evolution of the character of God. Now this is found in the plan of our redemption; nor do we see how, in any transaction between the great

Father of existence, and the children who have sprung from Him, the moral attributes of the Deity could, if we may so express ourselves, be put to so severe and so delicate a test. It is true, that the great matters of sin and of salvation, fall without impression on the heavy ears of a listless and alienated world. But they who, to use the language of the Bible, are light in the Lord, look otherwise at these things. They see sin in all its malignity, and salvation in all its mysterious greatness. And it would put them on the stretch of all their faculties, when they saw rebellion lifting up its standard against the Majesty of heaven, and the truth and the justice of God embarked on the threatenings He had uttered against all the doers of iniquity, and the honours of that august throne, which has the firm pillars of immutability to rest upon, linked with the fulfilment of the law that had come out from it; and when nothing else was looked for, but that God, by putting forth the power of His wrath, should accomplish His every denunciation, and vindicate the inflexibility of His government, and, by one sweeping deed of vengeance, assert, in the sight of all His creatures, the sovereignty which belonged to Him—with what desire must they have pondered on His ways, when, amid the urgency of all those demands which looked so high and so indispensable, they saw the unfoldings of the attribute of mercy—and that the Supreme Lawgiver was bending upon His guilty creatures an eye of tenderness—and that, in His profound and unsearchable wisdom, He was devising for them some plan of restoration—and that the eternal Son had to move from His dwelling-place in heaven, to carry it forward through all the difficulties by which it was encompassed—and that, after by the virtue of His mysterious sacrifice He had magnified the glory of every other perfection, He made mercy rejoice over them all, and threw open a way by which we sinful and polluted wanderers might, with the whole lustre of the Divine character untarnished, be re-admitted into fellowship with God, and be again brought back within the circle of His loyal and affectionate family.

Now, the essential character of such a transaction, viewed as a manifestation of God, does not hang upon the number of worlds, over which this sin and this salvation may have extended. We know that over this one world such an economy of wisdom and of mercy is instituted—and, even should this be the only world that is embraced by it, the moral display of the Godhead is mainly and substantially the same, as if it reached throughout the whole of that habitable extent which the science of astronomy has made known to us. By the disobedience of this one world, the law was trampled on—and, in the business of making truth and mercy to meet, and have a harmonious accomplishment on the men of this world, the dignity of God was put to the same trial; the justice of God appeared to lay the same immoveable barrier; the wisdom of God had to clear a way through the same difficulties; the forgiveness of God had to find the same mysterious conveyance to the sinners of a solitary world, as to the sinners of half a universe. The extent of the field upon which this question was decided, has no more influence on the question itself, than the figure or the dimensions of that field of combat, on which some great political question was fought, has on the importance or on the moral principles of the controversy that gave rise to it. This objection about the narrowness of the theatre, carries along with it all the grossness of materialism. To the eye of spiritual and intelligent beings, it is nothing. In their view, the redemption of a sinful world derives its chief interest from the display it gives of the mind and purposes of the Deity—and, should that world be but a single speck in the immensity of the works of God, the only way in which this affects their estimate of Him is to magnify His loving-kindness—who, rather than lose one solitary world of the myriads He has formed, would lavish all the riches of His beneficence and of His wisdom on the recovery of its guilty population.

Now, though it must be admitted that the Bible does not speak clearly or decisively as to the proper effect of redemption being extended to other worlds; it speaks most clearly and most decisively about the knowledge of its being disseminated amongst other orders of created intelligence than our own. But if the contemplation of God be their supreme enjoyment, then the very circumstance of our redemption being known to them, may invest it, even though it be but the redemption of one solitary world, with an importance as wide as the universe itself. It may spread amongst the hosts of immensity a new illustration of the character of Him who is all their praise; and in looking towards whom every energy within them is moved to the exercise of a deep and delighted admiration. The scene of the transaction may be narrow in point of material extent; while in the transaction itself there may be such a moral dignity, as to blazon the perfections of the Godhead over the face of creation; and, from the manifested glory of the Eternal, to send forth a tide of ecstasy, and of high gratulation, throughout the whole extent of His dependent provinces.

We shall not, in proof of the position, that the history of our redemption is known in other and distant places of creation, and is matter of deep interest and feeling amongst other orders of created intelligence—we shall not put down all the quotations which might be assembled together upon this argument. It is an impressive circumstance, that when Moses and Elias made a visit to our Saviour on the mount of transfiguration, and appeared in glory from heaven, the topic they brought along with them, and with which they were fraught, was the decease He was going to accomplish at Jerusalem. And however insipid the things of our salvation may be to an earthly understanding; we are made to know, that in the sufferings of Christ, and the glory which should follow, there is matter to attract the notice of celestial spirits, for these are the very things, says the Bible, which the angels desire to look into. And however listlessly we, the dull and grovelling children of an exiled family, may feel about the perfections of the Godhead, and the display of these perfections in the economy of the Gospel; it is intimated to us in the book of God's message, that the creation has its districts and its provinces; and we accordingly read of thrones and dominions, and principalities and powers—and whether these terms denote the separate regions of government, or the beings who, by a commission granted from the sanctuary of heaven, sit in delegated authority over them—even in their eyes the mystery of Christ stands arrayed in all the splendour of unsearchable riches; for we are told that this mystery was revealed for the very intent, that unto the principalities and powers, in heavenly places, might be made known by the church, the manifold wisdom of God. And while we, whose prospect reaches not beyond the narrow limits of the corner we occupy, look on the dealings of God in the world, as carrying in them all the insignificancy of a provincial transaction; God Himself, whose eye reaches to places which our eye hath not seen, nor our ear heard of, neither hath it entered into the imagination of our heart to conceive, stamps a universality on the whole matter of the Christian salvation, by such revelations as the following:—That he is to gather together in one all things in Christ, both which are in heaven, and which are in earth, even in him—and that at the name of Jesus every knee should bow, of things in heaven, and things in earth, and things under the earth—and that by him God reconciled all things unto himself, whether they be things in earth, or things in heaven.

We will not say in how far some of these passages extend the proper effect of that redemption which is by Christ Jesus, to other quarters of the universe of God; but they at least go to establish a widely disseminated knowledge of this transaction amongst the other orders of created intelligence. And they give us a distant glimpse of something more extended. They present a faint opening, through which may be seen some few traces of a wider and a nobler dispensation. They bring before us a dim transparency, on the other side of which the images of an obscure magnificence dazzle indistinctly upon the eye; and tell us, that in the economy of redemption, there is a grandeur commensurate to all that is known of the other works and purposes of the Eternal. They offer us no details; and man, who ought not to attempt a wisdom above that which is written, should never put forth his hand to the drapery of that impenetrable curtain which God, in His mysterious wisdom, has spread over those ways, of which it is but a very small portion that we know of them. But certain it is, that we know so much of them from the Bible; and the Infidel, with all the pride of his boasted astronomy, knows so little of them, from any power of observation—that the baseless argument of his, on which we have dwelt so long, is overborne in the light of all that positive evidence which God has poured around the record of His own testimony, and even in the light of its more obscure and casual intimations.

The minute and variegated details of the way in which this wondrous economy is extended, God has chosen to withhold from us; but He has oftener than once, made to us a broad and a general announcement of its dignity. He does not tell us, whether the fountain opened in the house of Judah, for sin and for uncleanness, sends forth its healing streams to other worlds than our own. He does not tell us the extent of the atonement.

But He tells us that the atonement itself, known, as it is, among the myriads of the celestial, forms the high song of eternity; that the Lamb who was slain, is surrounded by the acclamations of one wide and universal empire; that the might of His wondrous achievements, spreads a tide of gratulation over the multitudes who are about His throne; and that there never ceases to ascend from the worshippers of Him, who washed us from our sins in his blood, a voice loud as from numbers without number, sweet as from blessed voices uttering joy, when heaven rings jubilee, and loud hosannahs fill the eternal regions.

"And I beheld, and I heard the voice of many angels round about the throne; and the number of them was ten thousand times ten thousand, and thousands of thousands; saying with a loud voice, Worthy is the Lamb that was slain to receive power, and riches, and wisdom, and strength, and honour, and glory, and blessing. And every creature which is in heaven, and on earth, and under the earth, and such as are in the sea, and all that are in them, heard I saying, Blessing, and honour, and glory, and power, be unto him that sitteth on the throne, and unto the Lamb, for ever and ever."

A king might have the whole of his reign crowded with the enterprises of glory; and by the might of his arms, and the wisdom of his counsels, might win the first reputation among the potentates of the world; and be idolized throughout all his provinces, for the wealth and the security that he had spread around them—and still it is conceivable, that by the act of a single day in behalf of a single family; by some soothing visitation of tenderness to a poor and solitary cottage; by some deed of compassion, which conferred enlargement and relief on one despairing sufferer; by some graceful movement of sensibility at a tale of wretchedness; by some noble effort of self-denial, in virtue of which he subdued his every purpose of revenge, and spread the mantle of a generous oblivion over the fault of the man who had insulted and aggrieved him; above all, by an exercise of pardon so skilfully administered, as that, instead of bringing him down to a state of defencelessness against the provocation of future injuries, it threw a deeper sacredness over him, and stamped a more inviolable dignity than ever on his person and character:—why, on the strength of one such performance, done in a single hour, and reaching no farther in its immediate effects than to one house, or to one individual, it is a most possible thing, that the highest monarch upon earth might draw such a lustre around him, as would eclipse the renown of all his public achievements—and that such a display of magnanimity, or of worth, beaming from the secrecy of his familiar moments, might waken a more cordial veneration in every bosom, than all the splendour of his conspicuous history—and that it might pass down to posterity as a more enduring monument of greatness, and raise him farther, by its moral elevation, above the level of ordinary praise; and when he passes in review before the men of distant ages, may this deed of modest, gentle, unobtrusive virtue, be at all times appealed to, as the most sublime and touching memorial of his name.

In like manner did the King eternal, immortal, and invisible, surrounded as He is with the splendours of a wide and everlasting monarchy, turn Him to our humble habitation; and the footsteps of God manifest in the flesh, have been on the narrow spot of ground we occupy; and small though our mansion be, amid the orbs and the systems of immensity, hither hath the King of glory bent His mysterious way, and entered the tabernacle of men, and in the disguise of a servant did he sojourn for years under the roof which canopies our obscure and solitary world. Yes, it is but a twinkling atom in the peopled infinity of worlds that are around it —but look to the moral grandeur of the transaction, and not to the material extent of the field upon which it was executed—and from the retirement of our dwelling-place, there may issue forth such a display of the Godhead, as will circulate the glories of His name amongst all his worshippers. Here sin entered. Here was the kind and unwearied beneficence of a Father, repaid by the ingratitude of a whole family. Here the law of God was dishonoured, and that too in the face of its proclaimed and unalterable sanctions. Here the mighty contest of the attributes was ended—and when justice put forth its demands, and truth called for the fulfilment of its warnings, and the immutability of God would not recede by a single iota from any one of its positions, and all the severities He ever uttered against the children of iniquity, seemed to gather into one cloud of threatening vengeance on the tenement that held us—did the visit of the only-begotten Son chase away all these obstacles to the triumph of mercy—and humble as the tenement may be, deeply shaded in the obscurity of insignificance as it is, among the statelier mansions which are on every side of it—yet will the recall of its exiled family never be forgotten, and the illustration that has been given here of the mingled grace and majesty of God, will never lose its place among the themes and the acclamations of eternity.

And here it may be remarked, that as the earthly king who throws a moral aggrandizement around him by the act of a single day, finds, that after its performance he may have the space of many years for gathering to himself the triumphs of an extended reign—so the King who sits on high, and with whom one day is as a thousand years, and a thousand years as one day, will find, that after the period of that special administration is ended, by which this strayed world is again brought back within the limits of His favoured creation, there is room enough along the mighty track of eternity, for accumulating upon Himself a glory as wide and as universal as is the extent of his dominions. You will allow the most illustrious of this world's potentates, to give some hour of his private history to a deed of cottage or of domestic tenderness; and every time you think of the interesting story, you will feel how sweetly and how gracefully the remembrance of it blends itself with the fame of his public achievements. But still you think that there would not have been room enough for these achievements of his, had much of his time been spent, either amongst the habitations of the poor, or in the retirement of his own family; and you conceive, that it is because a single day bears so small a proportion to the time of his whole history, that he has been able to combine an interesting display of private worth, with all that brilliancy of exhibition, which has brought him down to posterity in the character of an august and a mighty sovereign.

Now apply this to the matter before us. Had the history of our redemption been confined within the limits of a single day, the argument that Infidelity has drawn from the multitude of other worlds would never have been offered. It is true, that ours is but an insignificant portion of the territory of God —but if the attentions by which He has signalized it, had only taken up a single day, this would never have occurred to us as forming any sensible withdrawment of the mind of the Deity from the concerns of His vast and universal government. It is the time which the plan of our salvation requires, that startles all those on whom this argument has any impression. It is the time taken up about this paltry world, which they feel to be out of proportion to the number of other worlds, and to the immensity of the surrounding creation. Now, to meet this impression, we do not insist at present on what we have already brought forward, that God, whose ways are not as our ways, can have His eye at the same instant on every place, and can divide and diversify His attention into any number of distinct exercises. What we have now to remark is, that the Infidel who urges the astronomical objection to the truth of Christianity, is only looking with half an eye to the principle on which it rests. Carry out the principle, and the objection vanishes.

He looks abroad on the immensity of space, and tells us how impossible it is, that this narrow corner of it can be so distinguished by the attentions of the Deity. Why does he not also look abroad on the magnificence of eternity; and perceive how the whole period of these peculiar attentions, how the whole time which elapses between the fall of man and the consummation of the scheme of his recovery, is but the twinkling of a moment to the mighty roll of innumerable ages? The whole interval between the time of Jesus Christ's leaving his Father's abode to sojourn amongst us, to that time when He shall have put all his enemies under His feet, and delivered up the kingdom to God even His Father, that God may be all in all; the whole of this interval bears as small a proportion to the whole of the Almighty's reign, as this solitary world does to the universe around it; and an infinitely smaller proportion than any time, however short, which an earthly monarch spends on some enterprise of private benevolence, does to the whole walk of his public and recorded history.

Why then does not the man, who can shoot his conceptions so sublimely abroad over the field of an immensity that knows no limits—why does he not also shoot them forward through the vista of a succession, that ever flows without stop and without termination? He has stept across the confines of this world's habitation in space, and out of the field which lies on the other side of it has he gathered an argument against the truth of revelation. We feel that we have nothing to do but to step across the confines of this world's history in time, and out of the futurity which lies beyond it can we gather that which will blow the argument to pieces, or stamp upon it all the narrowness of a partial and mistaken calculation. The day is coming when the whole of this wondrous history shall be looked back upon by the eye of remembrance, and be regarded as one incident in the extended annals of creation; and, with all the illustration and all the glory it has thrown on the character of the Deity, will it be seen as a single step in the evolution of His designs; and long as the time may appear, from the first act of our redemption to its final accomplishment, and close and exclusive as we may think the attentions of God upon it, it will be found that it has left Him room enough for all His concerns; and that, on the high scale of eternity, it is but one of those passing and ephemeral transactions which crowd the history of a never-ending administration.

DISCOURSE V.

ON THE SYMPATHY THAT IS FELT FOR MAN IN THE DISTANT PLACES OF CREATION.

"I say unto you, That likewise joy shall be in heaven over one sinner that repenteth, more than over ninety and nine just persons, which need no repentance."—LUKE xv. 7.

We have already attempted at full length to establish the position, that the infidel argument of astronomers goes to expunge a natural perfection from the character of God, even that wondrous property of His, by which He, at the same instant of time, can bend a close and a careful attention on a countless diversity of objects, and diffuse the intimacy of His power and of His presence, from the greatest to the minutest and most insignificant of them all. We also adverted shortly to this other circumstance, that it went to impair a moral attribute of the Deity. It goes to impair the benevolence of His nature. It is saying much for the benevolence of God, to say, that a single world, or a single system, is not enough for it—that it must have the spread of a mightier region, on which it may pour forth a tide of exuberancy throughout all its provinces—that as far as our vision can carry us, it has strewed immensity with the floating receptacles of life, and has stretched over each of them the garniture of such a sky as mantles our own habitation—and that even from distances which are far beyond the reach of human eye, the songs of gratitude and praise may now be arising to the one God, who sits surrounded by the regards of His one great and universal family.

Now it is saying much for the benevolence of God, to say, that it sends forth these wide and distant emanations over the surface of a territory so ample, that the world we inhabit, lying imbedded, as it does, amidst so much surrounding greatness, shrinks into a point that to the universal eye might appear to be almost imperceptible. But does it not add to the power and to the perfection of this universal eye, that at the very moment it is taking a comprehensive survey of the vast, it can fasten a steady and undistracted attention on each minute and separate portion of it; that at the very moment it is looking at all worlds, it can look most pointedly and most intelligently to each of them; that at the very moment it sweeps the field of immensity, it can settle all the earnestness of its regards upon every distinct handbreadth of that field; that at the very moment at which it embraces the totality of existence, it can send a most thorough and penetrating inspection into each of its details, and into every one of its endless diversities? We cannot fail to perceive how much this adds to the power of the all-seeing eye. Tell us then, if it do not add as much perfection to the benevolence of God, that while it is expatiating over the vast field of created things, there is not one portion of the field overlooked by it; that while it scatters blessings over the whole of an infinite range, it causes them to descend in a shower of plenty on every separate habitation; that while His arm is underneath and round about all worlds, He enters within the precincts of every one of them, and gives a care and a tenderness to each individual of their teeming population. Does not the God, who is said to be love, shed over this attribute of his its finest illustration—when, while He sits in the highest heaven, and pours out His fulness on the whole subordinate domain of nature and of providence, He bows a pitying regard on the very humblest of His children, and sends His reviving Spirit into every heart, and cheers by His presence every home, and provides for the wants of every family, and watches every sick-bed, and listens to the complaints of every sufferer; and while by his wondrous mind the weight of universal government is borne, is it not more wondrous and more excellent still, that He feels for every sorrow, and has an ear open to every prayer?

"It doth not yet appear what we shall be," says the apostle John, "but we know that when he shall appear, we shall be like him, for we shall see him as he is." It is the present lot of the angels, that they behold the face of our Father in heaven, and it would seem as if the effect of this was to form and to perpetuate in them the moral likeness of Him self, and that they reflect back upon Him His own image, and that thus a diffused resemblance to the Godhead is kept up amongst all those adoring worshippers who live in the near and rejoicing contemplation of the Godhead. Mark then how that peculiar and endearing feature in the goodness of the Deity, which we have just now adverted to—mark how beauteously it is reflected downwards upon us in the revealed attitude of angels. From the high eminences of heaven, are they bending a wakeful regard over the men of this sinful world; and the repentance of every one of them spreads a joy and a high gratulation throughout all its dwelling-places. Put this trait of the angelic character into contrast with the dark and louring spirit of an Infidel. He is told of the multitude of other worlds, and he feels a kindling magnificence in the conception, and he is seduced by an elevation which he cannot carry, and from this airy summit does he look down on the insignificance of the world we occupy, and pronounces it to be unworthy of those visits and of those attentions which we read of in the New Testament. He is unable to wing his upward way along the scale, either of moral or of natural perfection; and when the wonderful extent of the field is made known to him, over which the wealth of the Divinity is lavished—there he stops, and wilders, and altogether misses this essential perception, that the power and perfection of the Divinity are not more displayed by the mere magnitude of the field, than they are by that minute and exquisite filling up, which leaves not its smallest portions neglected; but which imprints the fulness of the Godhead upon every one of them; and proves, by every flower of the pathless desert, as well as by every orb of immensity, how this unsearchable Being can care for all, and provide for all, and, throned in mystery too high for us, can, throughout every instant of time, keep His attentive eye on every separate thing that He has formed, and, by an act of His thoughtful and presiding intelligence, can constantly embrace all.

But God, compassed about as He is with light inaccessible, and full of glory, lies so hidden from the ken and conception of all our faculties, that the spirit of man sinks exhausted by its attempts to comprehend Him. Could the image of the Supreme be placed direct before the eye of the mind, that flood of splendour, which is ever issuing from Him on all who have the privilege of beholding, would not only dazzle, but overpower us. And therefore it is, that we bid you look to the reflection of that image, and thus to take a view of its mitigated glories, and to gather the lineaments of the Godhead in the face of those righteous angels, who have never thrown away from them the resemblance in which they were created; and, unable as you are to support the grace and the majesty of that countenance, before which the seers and the prophets of other days fell, and became as dead men, let us, before we bring this argument to a close, borrow one lesson of Him who sitteth on the throne, from the aspect and the revealed doings of those who are surrounding it.

The Infidel, then, as he widens the field of his contemplations, would suffer its every separate object to die away into forgetfulness: these angels, expatiating as they do, over the range of a loftier universality, are represented as all awake to the history of each of its distinct and subordinate provinces. The Infidel, with his mind afloat among suns and among systems, can find no place in his already occupied regards, for that humble planet which lodges and accommodates our species: the angels, standing on a loftier summit, and with a mightier prospect of creation before them, are yet represented as looking down on this single world, and attentively marking the every feeling and the every demand of all its families. The Infidel, by sinking us down to an unnoticeable minuteness, would lose sight of our dwelling-place altogether, and spread a darkening shroud of oblivion over all the concerns and all the interests of men: but the angels will not so abandon us; and undazzled by the whole surpassing grandeur of that scenery which is around them, are they revealed as directing all the fulness of their regard to this our habitation, and casting a longing and a benignant eye on ourselves and on our children. The Infidel will tell us of those worlds which roll afar, and the number of which outstrips the arithmetic of the human understanding—and then, with the hardness of an unfeeling calculation, will he consign the one we occupy, with all its guilty generations, to despair. But He who counts the number of the stars, is set forth to us as looking at every inhabitant among the millions of our species, and by the word of the Gospel beckoning to him with the hand of invitation, and on the very first step of his return, as moving towards him with all the eagerness of the prodigal's father, to receive him back again into that presence from which he had wandered. And as to this world, in favour of which the scowling Infidel will not permit one solitary movement, all heaven is represented as in a stir about its restoration; and there cannot a single son, or a single daughter, be recalled from sin unto righteousness, without an acclamation of joy amongst the hosts of Paradise. And we can say it of the humblest and the unworthiest of you all, that the eye of angels is upon him, and that his repentance would, at this moment, send forth a wave of delighted sensibility throughout the mighty throng of their innumerable legions.

Now, the single question we have to ask, is, On which of the two sides of this contrast do we see most of the impress of heaven? Which of the two would be most glorifying to God? Which of them carries upon it most of that evidence which lies in its having a celestial character? For if it be the side of the Infidel, then must all our hopes expire with the ratifying of that fatal sentence, by which the world is doomed, through its insignificancy, to perpetual exclusion from the attentions of the Godhead. We have long been knocking at the door of your understanding, and have tried to find an admittance to it for many an argument. We now make our appeal to the sensibilities of your heart; and tell us to whom does the moral feeling within it yield its readiest testimony—to the Infidel, who would make this world of ours vanish away into abandonment—or to those angels, who ring throughout all their mansions the hosannas of joy, over every one individual of its repentant population?

And here we cannot omit to take advantage of that opening with which the Saviour has furnished us, by the parables of this chapter, and admits us into a familiar view of that principle on which the inhabitants of heaven are so awake to the deliverance and the restoration of our species. To illustrate the difference in the reach of knowledge and of affection, between a man and an angel, let us think of the difference of reach between one man and another. You may often witness a man, who feels neither tenderness nor care beyond the precincts of his own family; but who, on the strength of those instinctive fondnesses which nature has implanted in his bosom, may earn the character of an amiable father, or a kind husband, or a bright example of all that is soft and endearing in the relations of domestic society. Now conceive him, in addition to all this, to carry his affections abroad, without, at the same time, any abatement of their intensity towards the objects which are at home—that, stepping across the limits of the house he occupies, he takes an interest in the families which are near him—that he lends his services to the town or the district wherein he is placed, and gives up a portion of his time to the thoughtful labours of a humane and public-spirited citizen. By this enlargement in the sphere of his attention, he has extended his reach; and, provided he has not done so at the expense of that regard which is due to his family, a thing which, cramped and confined as we are, we are very apt, in the exercise of our humble faculties, to do—I put it to you, whether by extending the reach of his views and his affections, he has not extended his worth and his moral respectability along with it?

But we can conceive a still farther enlargement. We can figure to ourselves a man, whose wakeful sympathy overflows the field of his own immediate neighbourhood—to whom the name of country comes with all the omnipotence of a charm upon his heart, and with all the urgency of a most righteous and resistless claim upon his services—who never hears the name of Britain sounded in his ears, but it stirs up all his enthusiasm in behalf of the worth and the welfare of its people—who gives himself up, with all the devotedness of a passion, to the best and the purest objects of patriotism—and who, spurning away from him the vulgarities of party ambition, separates his life and his labours to the fine pursuit of augmenting the science, or the virtue, or the substantial prosperity of his nation. O! could such a man retain all the tenderness, and fulfil all the duties which home and which neighbourhood require of him, and at the same time, expatiate in the might of his untired faculties, on so wide a field of benevolent contemplation—would not this extension of reach place him still higher than before, on the scale both of moral and intellectual gradation, and give him a still brighter and more enduring name in the records of human excellence?

And, lastly, we can conceive a still loftier flight of humanity—a man, the aspiring of whose heart for the good of man, knows no limitations—whose longings and whose conceptions on this subject, overleap all the barriers of geography—who, looking on himself as a brother of the species, links every spare energy which belongs to him, with the cause of its amelioration—who can embrace within the grasp of his ample desires, the whole family of mankind—and who, in obedience to a heaven-born movement of principle within him, separates himself to some big and busy enterprise, which is to tell on the moral destinies of the world. Could such a man mix up the softenings of private virtue, with the habit of so sublime a comprehension —if, amid those magnificent darings of thought and of performance, the mildness of his benignant eye could still continue to cheer the retreat of his family, and to spread the charm and the sacredness of piety among all its members—could he even mingle himself in all the gentleness of a soothed and a smiling heart, with the playfulness of his children —and also find strength to shed the blessings of his presence and his counsel over the vicinity around him;—would not the combination of so much grace with so much loftiness, only serve the more to aggrandize him? Would not the one ingredient of a character so rare, go to illustrate and to magnify the other? And would not you pronounce him to be the fairest specimen of our nature, who could so call out all your tenderness, while he challenged and compelled all your veneration?

Nor can we proceed, at this point of our argument, without adverting to the way in which this last and this largest style of benevolence is exemplified in our own country—where the spirit of the Gospel has given to many of its enlightened disciples, the impulse of such a philanthropy, as carries abroad their wishes and their endeavours to the very outskirts of human population—a philanthropy, of which, if you asked the extent or the boundary of its field, we should answer in the language of inspiration, that the field is the world —a philanthropy, which overlooks all the distinctions of cast and of colour, and spreads its ample regards over the whole brotherhood of the species—a philanthropy, which attaches itself to man in the general; to man throughout all his varieties; to man as the partaker of one common nature, and who, in whatever clime or latitude you may meet with him, is found to breathe the same sympathies, and to possess the same high capabilities both of bliss and of improvement. It is true, that, upon this subject, there is often a loose and unsettled magnificence of thought, which is fruitful of nothing but empty speculation. But the men to whom we allude, have not imaged the enterprise in the form of a thing unknown. They have given it a local habitation. They have bodied it forth in deed and in accomplishment. They have turned the dream into a reality. In them, the power of a lofty generalization meets with its happiest attemperment, in the principle and perseverance, and all the chastening and subduing virtues of the New Testament. And, were we in search of that fine union of grace and of greatness which we have now been insisting on, and in virtue of which, the enlightened Christian can at once find room in his bosom for the concerns of universal humanity, and for the play of kindliness towards every individual he meets with—we could no where more readily expect to find it, than with the worthies of our own land—the Howard of a former generation, who paced over Europe in quest of the unseen wretchedness which abounds in it—or in such men of our present generation, as Wilberforce, who lifted his unwearied voice against the biggest outrage ever practised on our nature, till he wrought its extermination—and Clarkson, who plied his assiduous task at rearing the materials of its impressive history, and, at length carried, for this righteous cause, the mind of Parliament—and Carey, from whose hand the generations of the East are now receiving the elements of their moral renovation—and, in fine, those holy and devoted men, who count not their lives dear unto them; but, going forth every year from the island of our habitation, carry the message of heaven over the face of the world; and, in the front of severest obloquy, are now labouring in remotest lands; and are reclaiming another and another portion from the wastes of dark and fallen humanity; and are widening the domains of gospel light and gospel principle amongst them; and are spreading a moral beauty around the every spot on which they pitched their lowly tabernacle; and are at length compelling even the eye and the testimony of gainsayers, by the success of their noble enterprise; and are forcing the exclamation of delighted surprise from the charmed and the arrested traveller, as he looks at the softening tints which they are now spreading over the wilderness, and as he hears the sound of the chapel bell, and as in those haunts where, at the distance of half a generation, savages would have scowled upon his path, he regales himself with the hum of missionary schools, and the lovely spectacle of peaceful and Christian villages.

Such, then, is the benevolence, at once so gentle

and so lofty, of those men, who, sanctified by the faith that is in Jesus, have had their hearts visited from heaven by a beam of warmth and of sacredness. What, then, we should like to know, is the benevolence of the place from whence such an influence cometh? How wide is the compass of this virtue there, and how exquisite is the feeling of its tenderness, and how pure and how fervent are its aspirings among those unfallen beings who have no darkness, and no encumbering weight of corruption to strive against? Angels have a mightier reach of contemplation. Angels can look upon this world and all which it inherits, as the part of a larger family. Angels were in the full exercise of their powers even at the first infancy of our species, and shared in the gratulations of that period, when, at the birth of humanity, all intelligent nature felt a gladdening impulse, and the morning stars sang together for joy. They loved us even with the love which a family on earth bears to a younger sister; and the very childhood of our tinier faculties did only serve the more to endear us to them; and though born at a later hour in the history of creation, did they regard us as heirs of the same destiny with themselves, to rise along with them in the scale of moral elevation, to bow at the same footstool, and to partake in those high dispensations of a parent's kindness and a parent's care, which are ever emanating from the throne of the Eternal on all the members of a duteous and affectionate family. Take the reach of an angel's mind, but, at the same time, take the seraphic fervour of an angel's benevolence along with it; how, from the eminence on which he stands, he may have an eye upon many worlds, and a remembrance upon the origin and the successive concerns of every one of them; how he may feel the full force of a most affecting relationship with the inhabitants of each, as the offspring of one common Father; and though it be both the effect and the evidence of our depravity, that we cannot sympathize with these pure and generous ardours of a celestial spirit; how it may consist with the lofty comprehension, and the ever-breathing love of an angel, that he can both shoot his benevolence abroad over a mighty expanse of planets and of systems, and lavish a flood of tenderness on each individual of their teeming population.

Keep all this in view, and you cannot fail to perceive how the principle, so finely and so copiously illustrated in this chapter, may be brought to meet the infidelity we have thus long been employed in combating. It was nature, and the experience of every bosom will affirm it—it was nature in the shepherd to leave the ninety and nine of his flock forgotten and alone in the wilderness, and betaking himself to the mountains, to give all his labour and all his concern to the pursuit of one solitary wanderer. It was nature—and we are told in the passage before us, that it is such a portion of nature as belongs not merely to men, but to angels—when the woman, with her mind in a state of listlessness as to the nine pieces of silver that were in secure custody, turned the whole force of her anxiety to the one piece which she had lost, and for which she had to light a candle, and to sweep the house, and to search diligently until she found it. It was nature in her to rejoice more over that piece than over all the rest of them, and to tell it abroad among friends and neighbours, that they might rejoice along with her—and sadly effaced as humanity is, in all her original lineaments, this is a part of our nature, the very movements of which are experienced in heaven, "where there is more joy over one sinner that repenteth, than over ninety and nine just persons who need no repentance." For any thing we know, every planet that rolls in the immensity around us may be a land of righteousness; and be a member of the household of God; and have her secure dwelling-place within that ample limit, which embraces His great and universal family. But we know at least of one wanderer; and how wofully she has strayed from peace and from purity; and how, in dreary alienation from Him who made her, she has bewildered herself amongst those many devious tracks, which have carried her afar from the path of immortality; and how sadly tarnished all those beauties and felicities are, which promised, on that morning of her existence when God looked on her, and saw that all was very good—which promised so richly to bless and to adorn her; and how, in the eye of the whole unfallen creation, she has renounced all this godliness, and is fast departing away from them into guilt, and wretchedness, and shame. If there be any truth in this chapter, and any sweet or touching nature in the principle which runs throughout all its parables, let us cease to wonder, though they who surround the throne of love should be looking so intently towards us—or though, in the way by which they have singled us out, all the other orbs of space should, for one short season, on the scale of eternity, appear to be forgotten—or though, for every step of her recovery, and for every individual who is rendered back again to the fold from which he was separated, another and another message of triumph should be made to circulate amongst the hosts of paradise—or though, lost as we are, and sunk in depravity as we are, all the sympathies of heaven should now be awake on the enterprise of Him who has travailed, in the greatness of his strength, to seek and to save us.

And here we cannot but remark how fine a harmony there is between the law of sympathetic nature in heaven, and the most touching exhibitions of it on the face of our world. When one of a numerous household droops under the power of disease, is not that the one to whom all the tenderness is turned, and who, in a manner, monopolises the inquiries of his neighbourhood, and the care of his family? When the sighing of the midnight storm sends a dismal foreboding into the mother's heart, to whom of all her offspring, we would ask, are her thoughts and her anxieties then wandering? Is it not to her sailor boy whom her fancy has placed amid the rude and angry surges of the ocean? Does not this, the hour of his apprehended danger, concentrate upon him the whole force of her wakeful meditations? And does not he engross, for a season, her every sensibility, and her every prayer? We sometimes hear of shipwrecked passengers thrown upon a barbarous shore; and seized upon by its prowling inhabitants; and hurried away through the tracks of a dreary and unknown wilderness; and sold into captivity; and loaded with the fetters of irrecoverable bondage; and who, stripped of every other liberty but the liberty of thought, feel even this to be another ingredient of wretchedness, for what can they think of but home? and, as all its kind and tender imagery comes upon their remembrance, how can they think of it but in the bitterness of despair? Oh tell us, when the fame of all this disaster reaches his family, who is the member of it to whom is directed the full tide of its griefs and of its sympathies? Who is it that, for weeks and for months, usurps their every feeling, and calls out their largest sacrifices, and sets them to the busiest expedients for getting him back again? Who is it that makes them forgetful of themselves and of all around them? and tell us if you can assign a limit to the pains, and the exertions, and the surrenders which afflicted parents and weeping sisters would make to seek and to save him?

Now conceive, as we are warranted to do by the parables of this chapter, the principle of all these earthly exhibitions to be in full operation around the throne of God. Conceive the universe to be one secure and rejoicing family, and that this alienated world is the only strayed, or only captive member belonging to it; and we shall cease to wonder, that, from the first period of the captivity of our species, down to the consummation of their history in time, there should be such a movement in heaven; or that angels should so often have sped their commisioned way on the errand of our recovery; or that the Son of God should have bowed Himself down to the burden of our mysterious atonement; or that the Spirit of God should now, by the busy variety of His all-powerful influences, be carrying forward that dispensation of grace which is to make us meet for re-admittance into the mansions of the celestial. Only think of love as the reigning principle there; of love, as sending forth its energies and aspirations to the quarter where its object is most in danger of being for ever lost to it; of love, as called forth by this single circumstance to its uttermost exertion, and the most exquisite feeling of its tenderness; and then shall we come to a distinct and familiar explanation of this whole mystery: nor shall we resist, by our incredulity, the gospel message any longer, though it tells us, that throughout the whole of this world's history, long in our eyes, but only a little month in the high periods of immortality, so much of the vigilance, and so much of the earnestness of heaven, should have been expended on the recovery of its guilty population.

There is another touching trait of nature, which goes finely to heighten this principle, and still more forcibly to demonstrate its application to our present argument. So long as the dying child of David was alive, he was kept on the stretch of anxiety and of suffering with regard to it. When it expired, he arose and comforted himself. This narrative of King David is in harmony with all that we experience of our own movements and our own sensibilities. It is the power of uncertainty which gives them so active and so interesting a play in our bosoms; and which heightens all our regards to a tenfold pitch of feeling and of exercise; and which fixes down our watchfulness upon our infant's dying bed; and which keeps us so painfully alive to every turn and to every symptom in the progress of its malady; and which draws out all our affections for it to a degree of intensity that is quite unutterable; and which urges us on to ply our every effort and our every expedient, till hope withdraw its lingering beam, or till death shut the eyes of our beloved in the slumber of its long and its last repose.

We know not who of you have your names written in the book of life—nor can we tell if this be known to the angels which are in heaven. While in the land of living men, you are under the power and application of a remedy, which, if taken as the Gospel prescribes, will renovate the soul, and altogether prepare it for the bloom and the vigour of immortality. Wonder not then, that with this principle of uncertainty in such full operation, ministers should feel for you; or angels should feel for you; or all the sensibilities of heaven should be awake upon the symptoms of your grace and reformation; or the eyes of those who stand upon the high eminences of the celestial world, should be so earnestly fixed on every footstep and new evolution of your moral history. Such a consideration as this should do something more than silence the infidel objection. It should give a practical effect to the calls of repentance. How will it go to aggravate the whole guilt of our impenitency, should we stand out against the power and the tenderness of these manifold applications—the voice of a beseeching God upon us—the word of salvation at our very door—the free offer of strength and of acceptance sounded in our hearing—the Spirit in readiness with His agency to meet our every desire and our every inquiry—angels beckoning us to their company—and the very first movements of our awakened conscience, drawing upon us all their regards and all their earnestness!

DISCOURSE VI.

ON THE CONTEST FOR AN ASCENDANCY OVER MAN, AMONGST THE HIGHER ORDERS OF INTELLIGENCE.

"And, having spoiled principalities and powers, he made a show of them openly, triumphing over them in it."—COLOSSIANS ii. 15.

THOUGH these Discourses be now drawing to a close, it is not because we feel that much more might not be said on the subject of them, both in the way of argument and of illustration. The whole of the infidel difficulty proceeds upon the assumption, that the exclusive bearing of Christianity is upon the people of our earth; that this solitary planet is in no way implicated with the concerns of a wider dispensation; that the revelation we have of the dealings of God in this district of His empire, does not suit and subordinate itself to a system of moral administration, as extended as is the whole of his monarchy. Or, in other words, because Infidels have not access to the whole truth, do they refuse a part of it, however well attested or well accredited it may be; because a mantle of deep obscurity rests on the government of God, when taken in all its eternity and all its entireness, do they shut their eyes against that allowance of light which has been made to pass downwards upon our world from time to time, through so many partial unfoldings; and till they are made to know the share which other planets have in these communications of mercy, do they turn them away from the actual message which has come to their own door, and will neither examine its credentials, nor be alarmed by its warnings, nor be won by the tenderness of its invitations.

On that day when the secrets of all hearts shall be revealed, there will be found such a wilful duplicity and darkening of the mind in the whole of this proceeding, as shall bring down upon it the burden of a righteous condemnation. But even now does it lie open to the rebuke of philosophy, when the soundness and the consistency of her principles are brought faithfully to bear upon it. Were the character of modern science rightly understood, it would be seen, that the very thing which gave such strength and sureness to all her conclusions, was that humility of spirit which belonged to her. She promulgates all that is positively known; but she maintains the strictest silence and modesty about all that is unknown. She thankfully accepts of evidence wherever it can be found; nor does she spurn away from her the very humblest contribution of such doctrine, as can be witnessed by human observation, or can be attested by human veracity. But with all this she can hold out most sternly against that power of eloquence and fancy, which often throws so bewitching a charm over the plausibilities of ingenious speculation. Truth is the alone object of her reverence; and did she at all times keep by her attachments, nor throw them away when theology submitted to her cognizance its demonstrations and its claims, we should not despair of witnessing as great a revolution in those prevailing habitudes of thought which obtain throughout our literary establishments, on the subject of Christianity, as that which has actually taken place in the views which obtain on the philosophy of external nature. This is the first field on which have been successfully practised the experimental lessons of Bacon; and they who are conversant with these matters, know how great and how general a uniformity of doctrine now prevails in the sciences of astronomy, and mechanics, and chemistry, and almost all the other departments in the history and philosophy of matter. But this uniformity stands strikingly contrasted with the diversity of our moral systems, with the restless fluctuations both of language and of sentiment which are taking place in the philosophy of mind, with the palpable fact, that every new course of instruction upon this subject, has some new articles, or some new explanations to peculiarize it: and all this is to be attributed, not to the progress of the science, not to a growing, but to an alternating movement, not to its perpetual additions, but to its perpetual vibrations.

We mean not to assert the futility of moral science, or to deny her importance, or to insist on the utter hopelessness of her advancement. The Baconian method will not probably push forward her discoveries with such a rapidity, or to such an extent, as many of her sanguine disciples have anticipated. But if the spirit and the maxims of this philosophy were at all times proceeded upon, it would certainly check that rashness and variety of excogitation, in virtue of which it may almost be said, that every new course presents us with a new system, and that every new teacher has some singularity or other to characterize him. She may be able to make out an exact transcript of the phenomena of mind, and in so doing, she yields a most important contribution to the stock of human acquirements. But, when she attempts to grope her darkling way through the counsels of the Deity, and the futurities of His administration; when, without one passing acknowledgment to the embassy which professes to have come from him, or to the facts and to the testimonies by which it has so illustriously been vindicated, she launches forth her own speculations on the character of God, and the destiny of man; when, though this be a subject on which neither the recollections of history, nor the ephemeral experience of any single life, can furnish one observation to enlighten her, she will nevertheless utter her own plausibilities, not merely with a contemptuous neglect of the Bible, but in direct opposition to it; then it is high time to remind her of the difference between the reverie of him who has not seen God, and the well-accredited declaration of him who was in the beginning with God, and was God; and to tell her, that this, so far from being the argument of an ignoble fanaticism, is in harmony with the very argument upon which the science of experiment has been reared, and by which it has been at length delivered from the influence of theory, and purified of all its vain and visionary splendours.

In our last Discourses, we have attempted to collect, from the records of God's actual communication to the world, such traces of relationship between other orders of being and the great family of mankind, as serve to prove that Christianity is not so paltry and provincial a system as Infidelity presumes it to be. And as we said before, we have not exhausted all that may legitimately be derived upon this subject from the informations of Scripture. We have adverted, it is true, to the knowledge of our moral history which obtains throughout other provinces of the intelligent creation. We have asserted the universal importance which this may confer on the transactions even of one planet, in as much as it may spread an honourable display of the Godhead amongst all the mansions of infinity. We have attempted to expatiate on the argument, that an event little in itself, may be so pregnant with character, as to furnish all the worshippers of heaven with a theme of praise for eternity. We have stated that nothing is of magnitude in their eyes, but that which serves to endear to them the Father of their spirits, or to shed a lustre over the glory of His incomprehensible attributes—and that thus, from the redemption even of our solitary species, there may go forth such an exhibition of the Deity, as shall bear the triumphs of His name to the very outskirts of the universe.

We have farther adverted to another distinct Scriptural intimation, that the state of fallen man was not only matter of knowledge to other orders of creation, but was also matter of deep regret and affectionate sympathy; that agreeably to such laws of sympathy as are most familiar even to human observation, the very wretchedness of our condition was fitted to concentrate upon us the feelings, and the attentions, and the services of the celestial—to single us out for a time to the gaze of their most earnest and unceasing contemplation—to draw forth all that was kind and all that was tender within them—and just in proportion to the need and to the helplessness of us miserable exiles from the family of God, to multiply upon us the regards, and call out in our behalf the fond and eager exertions of those who had never wandered away from Him. This appears from the Bible to be the style of that benevolence which glows and which circulates around the throne of heaven. It is the very benevolence which emanates from the throne itself, and the attentions of which have for so many thousand years signalized the inhabitants of our world. This may look a long period for so paltry a world. But how have Infidels come to their conception that our world is so paltry? By looking abroad over the countless systems of immensity. But why then have they missed the conception, that the time of those peculiar visitations, which they look upon as so disproportionate to the magnitude of this earth, is just as evanescent as the earth itself is insignificant? Why look they not abroad on the countless generations of eternity; and thus come back to the conclusion, that after all, the redemption of our species is but an ephemeral doing in the history of intelligent nature; that it leaves the Author of it room for all the accomplishments of a wise and equal administration; and not to mention, that even during the progress of it, it withdraws not a single thought or a single energy of His, from other fields of creation, that there remains time enough to Him for carrying round the visitations of as striking and as peculiar a tenderness, over the whole extent of His great and universal monarchy?

It might serve still farther to incorporate the concerns of our planet with the general history of moral and intelligent beings, to state, not merely the knowledge which they take of us, and not merely the compassionate anxiety which they feel for us; but to state the importance derived to our world from its being the actual theatre of a keen and ambitious contest amongst the upper orders of creation. You know that for the possession of a very small and insulated territory, the mightiest empires of the world have put forth all their resources; and on some field of mustering competition, have monarchs met, and embarked for victory, all the pride of a country's rank, and all the flower and strength of a country's population. The solitary island around which so many fleets are hovering, and on the shores of which so many armed men are descending as to an arena of hostility, may well wonder at its own unlooked-for estimation. But other principles are animating the battle; and the glory of nations is at stake; and a much higher result is in the contemplation of each party, than the gain of so humble an acquirement as the primary object of the war; and honour, dearer to many a bosom than existence, is now the interest on which so much blood and so much treasure is expended; and the stirring spirit of emulation has now got hold of the combatants; and thus, amid all the insignificancy which attaches to the material origin of the contest, do both the eagerness and the extent of it, receive from the constitution of our nature, their most full and adequate explanation.

Now, if this be also the principle of higher natures —if, on the one hand, God be jealous of his honour; and, on the other, there be proud and exalted spirits who scowl defiance at Him and at His monarchy —if, on the side of heaven, there be an angelic host rallying around the standard of loyalty, who flee with alacrity at the bidding of the Almighty, who are devoted to His glory, and feel a rejoicing interest in the evolution of His counsels; and if, on the side of hell, there be a sullen front of resistance, a hate and malice inextinguishable, an unquelled daring of revenge to baffle the wisdom of the Eternal, and to arrest the hand, and to defeat the purposes of Omnipotence—then let the material prize of victory be insignificant as it may, it is the victory in itself which upholds the impulse of this keen and stimulated rivalry. If, by the sagacity of one infernal mind, a single planet has been seduced from its allegiance, and been brought under the ascendancy of him who is called in Scripture, "the god of this world;" and if the errand on which our Redeemer came, was to destroy the works of the devil—then let this planet have all the littleness which astronomy has assigned to it—call it what it is, one of the smaller islets which float on the ocean of vacancy; it has become the theatre of such a competition, as may have all the desires and all the energies of a divided universe embarked upon it. It involves in it other objects than the single recovery of our species. It decides higher questions. It stands linked with the supremacy of God, and will at length demonstrate the way in which He inflicts chastisement and overthrow upon all His enemies. We know not if our rebellious world be the only stronghold which Satan is possessed of, or if it be but the single post of an extended warfare, that is now going on between the powers of light and of darkness. But be it the one or the other, the parties are in array, and the spirit of the contest is in full energy, and the honour of mighty combatants is at stake; and let us therefore cease to wonder that our humble residence has been made the theatre of so busy an operation, or that the ambition of loftier natures has here put forth all its desire and all its strenuousness.

This unfolds to us another of those high and extensive bearings, which the moral history of our globe may have on the system of God's universal administration. Were an enemy to touch the shore of this high-minded country, and to occupy so much as one of the humblest of its villages, and there to seduce the natives from their loyalty, and to sit down along with them in entrenched defiance to all the threats, and to all the preparations of an insulted empire—how would the cry of wounded pride resound throughout all the ranks and varieties of our mighty population; and this very movement of indignancy would reach the king upon his throne; and circulate among those who stood in all the grandeur of chieftainship around him; and be heard to thrill in the eloquence of parliament; and spread so resistless an appeal to a nation's honour, and a nation's patriotism, that the trumpet of war would summon to its call all the spirit and all the willing energies of our kingdom; and rather than sit down in patient endurance under the burning disgrace of such a violation, would the whole of its strength and resources be embarked upon the contest; and never, never would we let down our exertions and our sacrifices, till either our deluded countrymen were reclaimed, or till the whole of this offence were, by one righteous act of vengeance, swept away altogether from the face of the territory it deformed.

The Bible is always most full and most explanatory on those points of revelation in which men are personally interested. But it does at times offer a dim transparency, through which may be caught a partial view of such designs and of such enterprises as are now afloat among the upper orders of intelligence. It tells us of a mighty struggle that is now going on for a moral ascendancy over the hearts of this world's population. It tells us that our race were seduced from their allegiance to God, by the plotting sagacity of one who stands pre-eminent against Him, among the hosts of a very wide and extended rebellion. It tells us of the Captain of salvation, who undertook to spoil him of this triumph; and throughout the whole of that magnificent train of prophecy which points to Him, does it describe the work he had to do, as a conflict, in which strength was to be put forth, and painful suffering to be endured, and fury to be poured upon enemies, and principalities to be dethroned, and all those toils, and dangers, and difficulties to be borne, which strewed the path of perseverance that was to carry him to victory.

But it is a contest of skill, as well as of strength and of influence. There is the earnest competition of angelic faculties embarked on this struggle for ascendancy. And while in the Bible there is recorded, (faintly and partially, we admit,) the deep and insidious policy that is practised on the one side; we are also told, that, on the plan of our world's restoration, there are lavished all the riches of an unsearchable wisdom upon the other. It would appear that, for the accomplishment of his purpose, the great enemy of God and of man plied his every calculation; and brought all the devices of his deep and settled malignity to bear upon our species; and thought, that could he involve us in sin, every attribute of the Divinity stood staked to the banishment of our race from beyond the limits of the empire of righteousness; and, thus did he practise his invasions on the moral territory of the unfallen; and, glorying in his success, did he fancy and feel that he had achieved a permanent separation between the God who sitteth in heaven, and one at least of the planetary mansions which He had reared.

The errand of the Saviour was to restore this sinful world, and have its people re-admitted within the circle of heaven's pure and righteous family. But in the government of heaven, as well as in the government of earth, there are certain principles which cannot be compromised; and certain maxims of administration which must never be departed from; and a certain character of majesty and of truth, on which the taint even of the slightest violation can never be permitted; and a certain authority which must be upheld by the immutability of all its sanctions, and the unerring fulfilment of all its wise and righteous proclamations. All this was in the mind of the archangel, and a gleam of malignant joy shot athwart him, as he conceived his project for hemming our unfortunate species within the bound of an irrecoverable dilemma; and as surely as sin and holiness could not enter into fellowship, so surely did he think, that if man were seduced to disobedience, would the truth, and the justice, and the immutability of God, lay their insurmountable barriers on the path of his future acceptance.

It was only in that plan of recovery of which Jesus Christ was the author and the finisher, that the great adversary of our species met with a wisdom which overmatched him. It is true, that he had reared, in the guilt to which he seduced us, a mighty obstacle in the way of this lofty undertaking. But when the grand expedient was announced, and the blood of that atonement, by which sinners are brought nigh, was willingly offered to be shed for us; and the eternal Son, to carry this mystery into accomplishment, assumed our nature—then was the prince of that mighty rebellion, in which the fate and the history of our world are so deeply implicated, in visible alarm for the safety of all his acquisitions:—nor can the record of this wondrous history carry forward its narrative, without furnishing some transient glimpses of a sublime and a superior warfare, in which, for the prize of a spiritual dominion over our species, we may dimly perceive the contest of loftiest talent, and all the designs of heaven in behalf of man, met at every point of their evolution, by the counterworkings of a rival strength and a rival sagacity.

We there read of a struggle which the Captain of our salvation had to sustain, when the lustre of the Godhead lay obscured, and the strength of its omnipotence was mysteriously weighed down under the infirmities of our nature—how Satan singled Him out, and dared Him to the combat of the wilderness—how all his wiles and all his influences were resisted—how he left our Saviour in all the triumphs of unsubdued loyalty—how the progress of this mighty achievement is marked by every character of a conflict—how many of the gospel miracles were so many direct infringements on the power and empire of a great spiritual rebellion—how, in one precious season of gladness among the few which brightened the dark career of our Saviour's humiliation, He rejoiced in spirit, and gave as the cause of it to his disciples, that "he saw Satan fall like lightning from heaven"—how the momentary advantages that were gotten over Him, are ascribed to the agency of this infernal being, who entered the heart of Judas, and tempted the disciple to betray His Master and His Friend. We know that we are treading on the confines of mystery. We cannot tell what the battle that he fought. We cannot compute the terror or the strength of his enemies. We cannot say, for we have not been told, how it was that they stood in marshalled and hideous array against Him:—nor can we measure how great the firm daring of His soul, when He tasted that cup in all its bitterness, which he prayed might pass away from Him; when, with the feeling that He was forsaken by His God, He trod the wine-press alone; when He entered single-handed upon that dreary period of agony, and insult, and death, in which, from the garden to the cross, He had to bear the burden of a world's atonement. We cannot speak in our own language, but we can say, in the language of the Bible, of the days and the nights of this great enterprise, that it was the season of the travail of His soul; that it was the hour and the power of darkness; that the work of our redemption, was a work accompanied by the effort, and the violence, and the fury of a combat; by all the arduousness of a battle in its progress, and all the glories of a victory in its termination: and after He called out that it was finished, after He was loosed from the prison-house of the grave, after He had ascended up on high, He is said to have made captivity captive; and to have spoiled principalities and powers; and to have seen His pleasure upon His enemies; and to have made a show of them openly.

We shall not affect a wisdom above that which is written, by fancying such details of this warfare as the Bible has not laid before us. But surely it is no more than being wise up to that which is written, to assert, that in achieving the redemption of our world, a warfare had to be accomplished; that upon this subject there was, among the higher provinces of creation, the keen and the animated conflict of opposing interests; that the result of it involved something grander and more affecting, than even the fate of this world's population; that it decided a question of rivalship between the righteous and everlasting Monarch of universal being, and the prince of a great and widely-extended rebellion, of which we neither know how vast is the magnitude, nor how important and diversified are the bearings: and thus do we gather, from this consideration, another distinct argument, helping us to explain why, on the salvation of our solitary species, so much attention appears to have been concentrated, and so much energy appears to have been expended.

But it would appear from the Records of Inspiration, that the contest is not yet ended; that on the one hand the Spirit of God is employed in making, for the truths of Christianity, a way into the human heart, with all the power of an effectual demonstration; that on the other, there is a spirit now abroad, which worketh in the children of disobedience: that on the one hand, the Holy Ghost is calling men out of darkness into the marvellous light of the Gospel; and that on the other hand, he who is styled the god of this world, is blinding their hearts, lest the light of the glorious Gospel of Christ should enter into them: that they who are under the dominion of the one, are said to have overcome, because greater is He that is in them than he that is in the world; and that they who are under the dominion of the other, are said to be the children of the devil, and to be under a snare, and to be taken captive by him at his will. How these respective powers do operate, is one question. The fact of their operation, is another. We abstain from the former. We attach ourselves to the latter, and gather from it, that the prince of darkness still walketh abroad amongst us; that he is still working his insidious policy, if not with the vigorous inspiration of hope, at least with the frantic energies of despair; that while the overtures of reconciliation are made to circulate through the world, he is plying all his devices to deafen and to extinguish the impression of them; or, in other words, while a process of invitation and of argument has emanated from heaven, for reclaiming men to their loyalty—the process is resisted at all its points, by one who is putting forth his every expedient, and wielding a mysterious ascendancy, to seduce and to enthrall them.

To an infidel ear, all this carries the sound of something wild and visionary along with it. But though only known through the medium of revelation; after it is known, who can fail to recognize its harmony with the great lineaments of human experience? Who has not felt the workings of a rivalry within him, between the power of conscience and the power of temptation? Who does not remember those seasons of retirement, when the calculations of eternity had gotten a momentary command over the heart; and time, with all its interests and all its vexations, had dwindled into insignificancy before them? And who does not remember, how, upon his actual engagement with the objects of time, they resumed a control, as great and as omnipotent, as if all the importance of eternity adhered to them—how they emitted from them such an impression upon his feelings, as to fix and to fascinate the whole man into a subserviency to their influence

—how, in spite of every lesson of their worthlessness, brought home to him at every turn by the rapidity of the seasons, and the vicissitudes of life, and the ever-moving progress of his own earthly career, and the visible ravages of death among his acquaintances around him, and the desolations of his family, and the constant breaking up of his system of friendships, and the affecting spectacle of all that lives and is in motion, withering and hastening to the grave;—how comes it, that, in the face of all this experience, the whole elevation of purpose, conceived in the hour of his better understanding, should be dissipated and forgotten? Whence the might, and whence the mystery of that spell, which so binds and so infatuates us to the world? What prompts us so to embark the whole strength of our eagerness and of our desires, in pursuit of interests which we know a few little years will bring to utter annihilation? Who is it that imparts to them all the charm and all the colour of an unfailing durability? Who is it that throws such an air of stability over these earthly tabernacles, as makes them look to the fascinated eye of man, like resting-places for eternity? Who is it that so pictures out the objects of sense, and so magnifies the range of their future enjoyment, and so dazzles the fond and deceived imagination, that, in looking onward through our earthly career, it appears like the vista, or the perspective, of innumerable ages? He who is called the god of this world. He who can dress the idleness of its waking dreams in the garb of reality. He who can pour a seducing brilliancy over the panorama of its fleeting pleasures and its vain anticipations. He who can turn it into an instrument of deceitfulness; and make it wield such an absolute ascendancy over all the affections, that man, become the poor slave of its idolatries and its charms, puts the authority of conscience and the warnings of the Word of God, and the offered instigations of the Spirit of God, and all the lessons of calculation, and all the wisdom even of his own sound and sober experience, away from him.

But this wondrous contest will come to a close. Some will return to their loyalty, and others will keep by their rebellion; and, in the day of the winding up of the drama of this world's history, there will be made manifest to the myriads of the various orders of creation, both the mercy and vindicated majesty of the Eternal. On that day, how vain will this presumption of the infidel astronomy appear, when the affairs of men come to be examined in the presence of an innumerable company; and beings of loftiest nature are seen to crowd around the judgment-seat; and the Saviour shall appear in our sky, with a celestial retinue, who have come with him from afar to witness all His doings, and to take a deep and solemn interest in all His dispensations; and the destiny of our species whom the Infidel would thus detach in solitary insignificance, from the universe altogether, shall be found to merge and to mingle with higher destinies —the good to spend their eternity with angels—the bad to spend their eternity with angels—the former to be re-admitted into the universal family of God's obedient worshippers—the latter to share in the everlasting pain and ignominy of the defeated hosts of the rebellious—the people of this planet to be implicated, throughout the whole train of their never-ending history, with the higher ranks, and the more extended tribes of intelligence: And thus it is, that the special administration we now live under, shall be seen to harmonize in its bearings, and to accord in its magnificence, with all that extent of nature and of her territories, which modern science has unfolded.

DISCOURSE VII.

ON THE SLENDER INFLUENCE OF MERE TASTE AND SENSIBILITY IN MATTERS OF RELIGION.

"And, lo! thou art unto them as a very lovely song of one that hath a pleasant voice, and can play well on an instrument; for they hear thy words, but they do them not."—EZEK. xxxiii. 32.

You easily understand how a taste for music is one thing, and a real submission to the influence of religion is another—how the ear may be regaled by the melody of sound, and the heart may utterly refuse the proper impression of the sense that is conveyed by it—how the sons and daughters of the world may, with their every affection devoted to its perishable vanities, inhale all the delights of enthusiasm, as they sit in crowded assemblage around the deep and solemn oratorio—and whether it be the humility of penitential feeling, or the rapture of grateful acknowledgment, or the sublime of a contemplative piety, or the aspiration of pure and of holy purposes, which breathes throughout the words of the performance, and gives to it all the spirit and all the expression by which it is pervaded; it is a very possible thing, that the moral, and the rational, and the active man, may have given no entrance into his bosom for any of these sentiments; and yet so overpowered may he be by the charm of the vocal conveyance through which they are addressed to him, that he may be made to feel with such an emotion, and to weep with such a tenderness, and to kindle with such a transport, and to glow with such an elevation, as may one and all carry upon them the semblance of sacredness.

But might not this semblance deceive him? Have you never heard any tell, and with complacency too, how powerfully his devotion was awakened by an act of attendance on the oratorio—how his heart, melted and subdued by the influence of harmony, did homage to all the religion of which it was the vehicle—how he was so moved and overborne, as to shed the tears of contrition, and to be agitated by the terrors of judgment, and to receive an awe upon his spirit of the greatness and the majesty of God—and that, wrought up to the lofty pitch of eternity, he could look down upon the world, and by the glance of one commanding survey, pronounce upon the littleness and the vanity of all its concerns? It is indeed very possible that all this might thrill upon the ears of the man, and circulate a succession of solemn and affecting images around his fancy—and yet that essential principle of his nature, upon which the practical influence of Christianity turns, might have met with no reaching and no subduing efficacy whatever to arouse it. He leaves the exhibition, as dead in trespasses and sins as he came to it. Conscience has not wakened upon him. Repentance has not turned him. Faith has not made any positive lodgement within him of her great and her constraining realities. He speeds him back to his business and to his family, and there he acts the old man in all the entireness of his uncrucified temper, and of his obstinate worldliness, and of all those earthly and unsanctified affections which are found to cleave to him with as great tenacity as ever. He is really and experimentally the very same man as before—and all those sensibilities which seemed to bear upon them so much of the air and unction of heaven, are found to go into dissipation, and be forgotten with the loveliness of the song.

Amid all that illusion which such momentary visitations of seriousness and of sentiment throw around the character of man, let us never lose sight of the test, that "by their fruits ye shall know them." It is not coming up to this test, that you hear and are delighted. It is that you hear and do. This is the ground upon which the reality of your religion is discriminated now; and on the day of reckoning, this is the ground upon which your religion will be judged then; and that award is to be passed upon you, which will fix and perpetuate your destiny for ever. You have a taste for music. This no more implies the hold and the ascendancy of religion over you, than that you have a taste for beautiful scenery, or a taste for painting, or even a taste for the sensualities of epicurism. But music may be made to express the glow and the movement of devotional feeling; and is it saying nothing to say that the heart of him who listens with a raptured ear, is, through the whole time of the performance, in harmony with such a movement? Why, it is saying nothing to the purpose. Music may lift the inspiring note of patriotism: and the inspiration may be felt; and it may thrill over the recesses of the soul, to the mustering up of all its energies; and it may sustain to the last cadence of the song, the firm nerve and purpose of intrepidity; and all this may be realized upon him, who, in the day of battle and upon actual collision with the dangers of it, turns out to be a coward. And music may lull the feelings into unison with piety; and stir up the inner man to lofty determinations; and so engage for a time his affections, that, as if weaned from the dust, they promise an immediate entrance on some great and elevated career, which may carry him through his pilgrimage superior to all the sordid and grovelling enticements that abound in it. But he turns him to the world, and all this glow abandons him; and the words which he had heard, he doeth them not; and in the hour of temptation he turns out to be a deserter from the law of allegiance; and the test we have now specified looks hard upon him, and discriminates him amid all the parading insignificance of his fine but fugitive emotions, to be the subject both of present guilt and of future vengeance.

The faithful application of this test would put to flight a host of other delusions. It may be carried round amongst all those phenomena of human character, where there is the exhibition of something associated with religion, but which is not religion itself. An exquisite relish for music is no test of the influence of Christianity. Neither are many other of the exquisite sensibilities of our nature. When a kind mother closes the eyes of her expiring babe, she is thrown into a flood of sensibility, and soothing to her heart are the sympathy and the prayers of an attending minister. When a gathering neighbourhood assemble to the funeral of an acquaintance, one pervading sense of regret and tenderness sits on the faces of the company; and the deep silence, broken only by the solemn utterance of the man of God, carries a kind of pleasing religiousness along with it. The sacredness of the hallowed day, and all the decencies of its observation, may engage the affections of him who loves to walk in the footsteps of his father; and every recurring Sabbath may bring to his bosom the charm of its regularity and its quietness. Religion has its accompaniments; and in these there may be a something to sooth and to fascinate, even in the absence of the appropriate influences of religion. The deep and tender impression of a family bereavement, is not religion. The love of established decencies, is not religion. The charm of all that sentimentalism which is associated with many of its solemn and affecting services, is not religion. They may form the distinct folds of its accustomed drapery; but they do not, any, or all of them put together, make up the substance of the thing itself. A mother's tenderness may flow most gracefully over the tomb of her departed little one; and she may talk the while of that heaven whither its spirit has ascended. The man whom death hath widowed of his friend, may abandon himself to the movements of that grief, which for a time will claim an ascendancy over him; and, amongst the multitude of his other reveries, may love to hear of the eternity, where sorrow and separation are alike unknown. He who has been trained from his infant days to remember the Sabbath, may love the holiness of its aspect, and associate himself with all its observances, and take a delighted share in the mechanism of its forms. But let not these think, because the tastes and the sensibilities which engross them, may be blended with religion, that they indicate either its strength or its existence within them. We recur to the test. We press its imperious exactions upon you. We call for fruit, and demand the permanency of a religious influence on the habits and the history. How many who take a flattering unction to their souls, when they think of their amiable feelings, and their becoming observations, with whom this severe touchstone would, like the head of Medusa, put to flight all their complacency! The afflictive dispensation is forgotten—and he on whom it was laid, is practically as indifferent to God and to eternity as before. The Sabbath services come to a close, and they are followed by the same routine of week-day worldliness as before. In neither the one case nor the other, do we see more of the radical influence of Christianity, than in the sublime and melting influence of sacred music upon the soul; and all this tide of emotion is found to die away from the bosom, like the pathos or like the loveliness of a song.

The instances may be multiplied without number. A man may have a taste for eloquence, and eloquence the most touching or sublime may lift her pleading voice on the side of religion. A man may love to have his understanding stimulated by the ingenuities, or the resistless urgencies of an argument; and argument the most profound and the most over-bearing, may put forth all the might of a constraining vehemence in behalf of religion. A man may feel the rejoicings of a conscious elevation, when some ideal scene of magnificence is laid before him; and where are these scenes so readily to be met with, as when led to expatiate in thought over the track of eternity, or to survey the wonders of creation, or to look to the magnitude of those great and universal interests which lie within the compass of religion? A man may have his attention riveted and regaled by that power of imitative description, which brings all the recollections of his own experience before him; which presents him with a faithful analysis of his own heart; which embodies in language such intimacies of observation and of feeling, as have often passed before his eyes, or played within his bosom, but had never been so truly or so ably pictured to the view of his remembrance. Now, all this may be done in the work of pressing the duties of religion; in the work of instancing the applications of religion; in the work of pointing those allusions to life and to manners, which manifest the truth to the conscience, and plant such a conviction of sin, as forms the very basis of a sinner's religion. Now, in all these cases, we see other principles brought into action, and which may be in a state of most lively and vigorous movement, and be yet in a state of entire separation from the principle of religion. We will venture to say, on the strength of these illustrations, that as much delight may emanate from the pulpit on an arrested audience beneath it, as ever emanated from the boards of a theatre—and with as total a disjunction of mind too, in the one case as in the other, from the essence or the habit of religion. We recur to the test. We make our appeal to experience; and we put it to you all, whether your finding upon the subject do not agree with our saying about it, that a man may weep and admire, and have many of his faculties put upon the stretch of their most intense gratification—his judgment established, and his fancy enlivened, and his feelings overpowered, and his hearing charmed as by the accents of heavenly persuasion, and all within him feasted by the rich and varied luxuries of an intellectual banquet!—Oh! it is cruel to frown unmannerly in the midst of so much satisfaction. But I must not forget that truth has her authority, as well as her sternness; and she forces me to affirm, that after all this has been felt and gone through, there might not be one principle which lies at the turning-point of conversion, that has experienced a single movement—not one of its purposes be conceived—not one of its doings be accomplished—not one step of that repentance, which, if we have not, we perish, so much as entered upon—not one announcement of that faith, by which we are saved, admitted into a real and actual possession by the inner man. He has had his hour's entertainment, and willingly does he award this homage to the performer, that he hath a pleasant voice, and can play well on an instrument—but, in another hour, it fleets away from his remembrance, and goes all to nothing, like the loveliness of a song

Now, in bringing these Discourses to a close,

we feel it our duty to advert to this exhibition of character in man. The sublime and interesting topic which has engaged us, however feebly it may have been handled; however inadequately it may have been put in all its worth, and in all its magnitude before you; however short the representation of the speaker, or the conception of the hearers, may have been of that richness, and that greatness, and that loftiness, which belong to it; possesses in itself a charm to fix the attention, and to regale the imagination, and to subdue the whole man into a delighted reverence; and, in a word, to beget such a solemnity of thought and of emotion, as may occupy and enlarge the soul for hours together, as may waft it away from the grossness of ordinary life, and raise it to a kind of elevated calm above all its vulgarities and all its vexations.

Now, tell us whether the whole of this effect upon the feelings may not be formed without the presence of religion. Tell us whether there might not be such a constitution of mind, that it may both want altogether that principle, in virtue of which the doctrines of Christianity are admitted into the belief, and the duties of Christianity are admitted into a government over the practice—and yet at the very same time, it may have the faculty of looking abroad over some scene of magnificence, and of being wrought up to ecstasy with the sense of all those glories among which it is expatiating. We want you to see clearly the distinction between these two attributes of the human character. They are, in truth, as different the one from the other, as a taste for the grand and the graceful of scenery differs from the appetite of hunger; and the one may both exist and have a most intense operation within the bosom of that very individual, who entirely disowns, and is entirely disgusted with the other. What! must a man be converted, ere from the most elevated peak of some Alpine wilderness, he become capable of feeling the force and the majesty of those great lineaments which the hand of nature has thrown around him, in the varied forms of precipice, and mountain, and the wave of mighty forests, and the rush of sounding waterfalls, and distant glimpses of human territory, and pinnacles of everlasting snow, and the sweep of that circling horizon, which folds in its ample embrace the whole of this noble amphitheatre? Tell us whether, without the aid of Christianity, or without a particle of reverence for the only Name given under heaven whereby men can be saved, a man may not kindle at such a perspective as this, into all the raptures, and into all the movements of a poetic elevation; and be able to render into the language of poetry, the whole of that sublime and beauteous imagery which adorns it? and, as if he were treading on the confines of a sanctuary which he has not entered, may he not mix up with the power and the enchantment of his description, such allusions to the presiding genius of the scene; or to the still but animating spirit of the solitude; or to the speaking silence of some mysterious character which reigns throughout the landscape; or, in fine, to that Eternal Spirit, who sits behind the elements He has formed, and combines them into all the varieties of a wide and a wondrous creation;—might not all this be said and sung with an emphasis so moving, as to spread the colouring of piety over the pages of him who performs thus well upon his instrument; and yet, the performer himself have a conscience unmoved by a single warning of God's actual communication, and the judgment unconvinced, and the fears unawakened, and the life unreformed by it?

Now, what is true of a scene on earth, is also true of that wider and more elevated scene which stretches over the immensity around it, into a dark and a distant unknown. Who does not feel an aggrandizement of thought and of faculty, when he looks abroad over the amplitudes of creation—when, placed on a telescopic eminence, his aided eye can find a pathway to innumerable worlds—when that wondrous field, over which there had hung for many ages the mantle of so deep an obscurity, is laid open to him, and, instead of a dreary and unpeopled solitude, he can see over the whole face of it such an extended garniture of rich and goodly habitations? Even the Atheist, who tells us that the universe is self-existent and indestructible—even he, who instead of seeing the traces of a manifold wisdom in its manifold varieties, sees nothing in them all but the exquisite structures and the lofty dimensions of materialism—even he, who would despoil creation of its God, cannot look upon its golden suns, and their accompanying systems, without the solemn impression of a magnificence that fixes and overpowers him. Now, conceive such a belief of God as you all profess, to dawn upon his understanding. Let him become as one of yourselves—and so be put into the condition of rising from the sublime of matter to the sublime of mind. Let him now learn to subordinate the whole of this mechanism to the design and authority of a great presiding Intelligence: and re-assembling all the members of the universe, however distant, into one family, let him mingle with his former conceptions of the grandeur which belong to it, the conception of that Eternal Spirit who sits enthroned on the immensity of His own wonders, and embraces all that He has made, within the ample scope of one great administration. Then will the images and the impressions of sublimity come in upon him from a new quarter. Then will another avenue be opened, through which a sense of grandeur may find its way into his soul, and have a mightier influence than ever to fill, and to elevate, and to expand it. Then will be established a new and a noble association, by the aid of which all that he formerly looked upon as fair, becomes more lovely; and all that he formerly looked upon as great, becomes more magnificent. But will you believe us, that even with this accession to his mind of ideas gathered from the contemplation of the Divinity; even with that pleasurable glow which steals over his imagination, when he now thinks of the majesty of God; even with as much of what you would call piety, as we fear is enough to sooth and to satisfy many of yourselves, and which stirs and kindles within you when you hear the goings forth of the Supreme set before you in the terms of a lofty representation; even with all this, we say, there may be as wide a distance from the habit and the character of godliness, as if God was still atheistically disowned by him. Take the conduct of his life and the currency of his affections; and you may see as little upon them of the stamp of loyalty to God, or of reverence for any one of his authenticated proclamations, as you may see in him who offers his poetic incense to the genii, or weeps enraptured over the visions of a beauteous mythology. The sublime of Deity has wrought up his soul to a pitch of conscious and pleasing elevation —and yet this no more argues the will of Deity to have a practical authority over him, than does that tone of elevation which is caught by looking at the sublime of a naked materialism. The one and the other have their little hour of ascendancy over him; and when he turns him to the rude and ordinary world, both vanish alike from his sensibilities, as does the loveliness of a song.

To kindle and be elevated by a sense of the majesty of God, is one thing. It is totally another thing, to feel a movement of obedience to the will of God, under the impression of His rightful authority over all the creatures whom He has formed. A man may have an imagination all alive to the former; while the latter never prompts him to one act of obedience; never leads him to compare his life with the requirements of the Lawgiver; never carries him from such a scrutiny as this, to the conviction of sin; never whispers such an accusation to the ear of his conscience, as causes him to mourn, and to be in heaviness for the guilt of his hourly and habitual rebellion; never shuts him up to the conclusion of the need of a Saviour; never humbles him to acquiescence in the doctrine of that revelation, which comes to his door with such a host of evidence, as even his own philosophy cannot bid away; never extorts a single believing prayer in the name of Christ, or points a single look, either of trust or of reverence, to His atonement; never stirs any effective movement of conversion; never sends an aspiring energy into his bosom after the aids of that Spirit, who alone can waken him out of his lethargies, and by the anointing which remaineth, can rivet and substantiate in his practice, those goodly emotions which have hitherto plied him with the deceitfulness of their momentary visits, and then capriciously abandoned him.

The mere majesty of God's power and greatness, when offered to your notice, lays hold of one of the faculties within you. The holiness of God, with His righteous claim of legislation, lays hold of another of these faculties. The difference between them is so great, that the one may be engrossed and interested to the full, while the other remains untouched, and in a state of entire dormancy. Now, it is no matter what it be that ministers delight to the former of these two faculties: If the latter be not arrested and put on its proper exercise, you are making no approximation whatever to the right habit and character of religion. There are a thousand ways in which we may contrive to regale your taste for that which is beauteous and majestic. It may find its gratification in the loveliness of a vale, or in the freer and bolder outlines of an upland situation, or in the terrors of a storm, or in the sublime contemplations of astronomy, or in the magnificent idea of a God who sends forth the wakefulness of His omniscient eye, and the vigour of His upholding hand, throughout all the realms of nature and of providence. The mere taste of the human mind may get its ample enjoyment in each and in all of these objects, or in a vivid representation of them; nor does it make any material difference, whether this representation be addressed to you from the stanzas of a poem, or from the recitations of a theatre, or finally from the discourses and the demonstrations of a pulpit. And thus it is, that still on the impulse of the one principle only, people may come in gathering multitudes to the house of God; and share with eagerness in all the glow and bustle of a crowded attendance; and have their every eye directed to the speaker; and feel a responding movement in their bosom to his many appeals and his many arguments; and carry a solemn and overpowering impression of all the services away with them; and yet, throughout the whole of this seemly exhibition, not one effectual knock may have been given at the door of conscience. The other principle may be as profoundly asleep, as if hushed into the insensibility of death. There is a spirit of deep slumber, it would appear, which the music of no description, even though attuned to a theme so lofty as the greatness and majesty of the Godhead, can ever charm away. Oh! it may have been a piece of parading insignificance altogether—the minister playing on his favourite instrument, and the people dissipating away their time on the charm and idle luxury of a theatrical emotion.

The religion of taste, is one thing. The religion of conscience, is another. We recur to the test. What is the plain and practical doing which ought to issue from the whole of our argument? If one lesson come more clearly or more authoritatively out of it than another, it is the supremacy of the Bible. If fitted to impress one movement rather than another; it is that movement of docility, in virtue of which, man, with the feeling that he has all to learn, places himself in the attitude of a little child, before the book of the unsearchable God, who has deigned to break His silence, and to transmit even to our age of the world, a faithful record of his own communication. What progress then are you making in this movement? Are you, or are you not, like new-born babes, desiring the sincere milk of the word, that you may grow thereby? How are you coming on in the work of casting down your lofty imaginations? With the modesty of true science, which is here at one with the humblest and most penitentiary feeling which Christianity can awaken, are you bending an eye of earnestness on the Bible, and appropriating its informations, and moulding your every conviction to its doctrines and its testimonies? How long, we beseech you, has this been your habitual exercise? By this time do you feel the darkness and the insufficiency of nature? Have you found your way to the need of an atonement? Have you learned the might and efficacy which are given to the principle of faith? Have you longed with all your energies to realize it? Have you broken loose from the obvious misdoings of your former history? Are you convinced of your total deficiency from the spiritual obedience of the affections? Have you read of the Holy Ghost, by whom renewed in the whole desire and character of your mind, you are led to run with alacrity in the way of the commandments? Have you turned to its practical use, the important truth, that He is given to the believing prayers of all, who really want to be relieved from the power both of secret and of visible iniquity? We demand something more than the homage you have rendered to the pleasantness of the voice that has been sounded in your hearing. What we have now to urge upon you, is the bidding of the voice, to read, and to reform, and to pray, and, in a word, to make your consistent step from the elevations of philosophy, to all those exercises, whether of doing or of believing, which mark the conduct of the earnest, and the devoted, and the subdued, and the aspiring Christian.

This brings under our view, a most deeply interesting exhibition of human nature, which may often be witnessed among the cultivated orders of society. When a teacher of Christianity addresses himself to that principle of justice within us, by which we feel the authority of God to be a prerogative which righteously belongs to Him, he is then speaking the appropriate language of religion, and is advancing its naked and appropriate claim over the obedience of mankind. He is then urging that pertinent and powerful consideration, upon which alone he can ever hope to obtain the ascendancy of a practical influence over the purposes and the conduct of human beings. It is only by insisting on the moral claim of God to a right of government over his creatures, that he can carry their loyal subordination to the will of God. Let him keep by this single argument, and urge it upon the conscience, and then, without any of the other accompaniments of what is called Christian oratory, he may bring convincingly home upon his hearers all the varieties of Christian doctrine. He may establish within their minds the dominion of all that is essential in the faith of the New Testament. He may, by carrying out this principle of God's authority into all its applications, convince them of sin. He may lead them to compare the loftiness and spirituality of His law, with the habitual obstinacy of their own worldly affections. He may awaken them to the need of a Saviour. He may urge them to a faithful and submissive perusal of God's own communication. He may thence press upon them the truth and the immutability of their Sovereign. He may work in their hearts an impression of this emphatic saying, that God is not to be mocked—that His law must be upheld in all the significancy of its proclamations—and that either its severities must be discharged upon the guilty, or in some other way an adequate provision be found for its outraged dignity, and its violated sanctions. Thus may he lead them to flee for refuge to the blood of the atonement. And he may further urge upon his hearers, that such is the enormity of sin, that it is not enough to have found an expiation for it; that its power and its existence must be eradicated from the hearts of all who are to spend their eternity in the mansions of the celestial; that for this purpose, an expedient is made known to us in the New Testament; that a process must be described upon earth, to which there is given the appropriate name of sanctification; that, at the very commencement of every true course of discipleship, this process is entered upon with a purpose in the mind of forsaking all; that nothing short of a single devotedness to the will of God, will ever carry us forward through the successive stages of this holy and elevated career; that to help the infirmities of our nature, the Spirit is ever in readiness to be given to those who ask it: and that thus the life of every Christian becomes a life of entire dedication to Him who died for us—a life of prayer and vigilance, and close dependence on the grace of God—and, as the infallible result of the plain but powerful and peculiar teaching of the Bible, a life of vigorous unwearied activity in the doing of all the commandments.

Now, this we should call the essential business of Christianity. This is the truth as it is in Jesus, in its naked and unassociated simplicity. In the work of urging it, nothing more might have been done, than to present certain views, which may come with as great clearness and freshness, and take as full possession of the mind of a peasant, as of the mind of a philosopher. There is a sense of God, and of the rightful allegiance that is due to Him. There are plain and practical appeals to the conscience. There is a comparison of the state of the heart, with the requirements of a law which proposes to take the heart under its obedience. There is the inward discernment of its coldness about God; of its unconcern about the matters of duty and of eternity; of its devotion to the forbidden objects of sense; of its constant tendency to nourish within its own receptacles, the very element and principle of rebellion, and in virtue of this, to send forth the stream of an hourly and accumulating disobedience over those doings of the outer man, which make up his visible history in the world. There is such an earnest and overpowering impression of all this, as will fix a man down to the single object of deliverance; as will make him awake only to those realities which have a significant and substantial bearing on the case that engrosses him; as will teach him to nauseate all the impertinences of tasteful and ambitious description; as will attach him to the truth in its simplicity; as will fasten his every regard upon the Bible, where, if he persevere in the work of honest inquiry, he will soon be made to perceive the accordancy between its statements, and all those movements of fear, or guilt, or deeply felt necessity, or conscious darkness, stupidity, and unconcern about the matters of salvation, which pass within his own bosom; in a word, as will endear to him that plainness of speech, by which his own experience is set evidently before him, and that plain phraseology of Scripture, which is best fitted to bring home to him the doctrine of redemption, in all the truth and in all the preciousness of its applications.

Now, the whole of this work may be going on, and that too in the wisest and most effectual manner, without so much as one particle of incense being offered to any of the subordinate principles of the human constitution. There may be no fascinations of style. There may be no magnificence of description. There may be no poignancy of acute and irresistible argument. There may be a riveted attention on the part of those whom the Spirit of God hath awakened to seriousness about the plain and affecting realities of conversion. Their conscience may be stricken, and their appetite be excited for an actual settlement of mind on those points about which they feel restless and unconfirmed. Such as these are vastly too much engrossed with the exigencies of their condition, to be repelled by the homeliness of unadorned truth. And thus it is, that while the loveliness of the song has done so little in helping on the influences of the gospel, our men of simplicity and prayer have done so much for it. With a deep and earnest impression of the truth themselves, they have made manifest that truth to the consciences of others. Missionaries have gone forth with no other preparation than the simple Word of the Testimony,—and thousands have owned its power, by being both the hearers of the word and the doers of it also. They have given us the experiment in a state of unmingled simplicity; and we learn, from the success of their noble example, that without any one human expedient to charm the ear, the heart may, by the naked instrumentality of the Word of God, urged with plainness on those who feel its deceit and its worthlessness, be charmed to an entire acquiescence in the revealed way of God, and have impressed upon it the genuine stamp and character of godliness.

Could the sense of what is due to God be effectually stirred up within the human bosom, it would lead to a practical carrying of all the lessons of Christianity. Now, to awaken this moral sense, there are certain simple relations between the creature and the Creator, which must be clearly apprehended, and manifested with power unto the conscience. We believe, that however much philosophers may talk about that comparative ease of forming those conceptions which are simple, they will, if in good earnest after a right footing with God, soon discover in their own minds, all that darkness and incapacity about spiritual things, which are so broadly announced to us in the New Testament. And oh! it is a deeply interesting spectacle, to behold a man, who can take a masterly and commanding survey over the field of some human speculation, who can clear his discriminated way through all the turns and ingenuities of some human argument, who, by the march of a mighty and resistless demonstration, can scale with assured footstep the sublimities of science, and, from his firm stand on the eminence he has won, can descry some wondrous range of natural or intellectual truth spread out in subordination before him:—and yet this very man, may, in reference to the moral and authoritative claims of the Godhead, be in a state of utter apathy and blindness! All his attempts, either at the spiritual discernment, or the practical impression of this doctrine, may be arrested and baffled by the weight of some great inexplicable impotency. A man of homely talents, and still homelier education, may see what he cannot see, and feel what he cannot feel; and wise and prudent as he is, there may lie the barrier of an obstinate and impenetrable concealment, between his accomplished mind, and those things which are revealed unto babes.

But while his mind is thus utterly devoid of what may be called the main or elemental principle of theology, he may have a far quicker apprehension, and have his taste and his feelings much more powerfully interested, than the simple Christian who is beside him, by what may be called the circumstantials of theology. He can throw a wider and more rapid glance over the magnitudes of creation. He can be more delicately alive to the beauties and the sublimities which abound in it. He can, when the idea of a presiding God is suggested to him, have a more kindling sense of His natural majesty, and be able, both in imagination and in words, to surround the throne of the Divinity by the blazonry of more great, and splendid, and elevating images. And yet, with all those powers of conception which he does possess, he may not possess that on which practical Christianity hinges. The moral relation between him and God, may neither be effectively perceived, nor faithfully proceeded on. Conscience may be in a state of the most entire dormancy, and the man be regaling himself with the magnificence of God, while he neither loves God, nor believes God, nor obeys God.

And here I cannot but remark, how much effect and simplicity go together in the annals of Moravianism. The men of this truly interesting denomination, address themselves exclusively to that principle of our nature, on which the proper influence of Christianity turns. Or, in other words, they take up the subject of the gospel message—that message devised by Him who knew what was in man, and who, therefore, knew how to make the right and the suitable application to man. They urge the plain Word of the Testimony: and they pray for a blessing from on high; and that thick impalpable veil, by which the god of this world blinds the hearts of them who believe not, lest the light of the glorious gospel of Christ should enter into them—that veil, which no power of philosophy can draw aside, gives way to the demonstration of the Spirit; and thus it is, that a clear perception of scriptural truth, and all the freshness and permanency of its moral influences, are to be met with among men who have just emerged from the rudest and the grossest barbarity. When one looks at the number and the greatness of their achievements—when he thinks on the change they have made on materials so coarse and so unpromising—when he eyes the villages they have formed —and around the whole of that engaging perspective by which they have chequered and relieved the grim solitude of the desert, he witnesses the love, and listens to the piety of reclaimed savages; —who would not long to be in possession of the charm by which they have wrought this wondrous transformation—who would not willingly exchange for it all the parade of human eloquence, and all the confidence of human argument—and for the wisdom of winning souls, who is there that would not rejoice to throw the loveliness of the song, and all the insignificancy of its passing fascinations away from him?

And yet it is right that every cavil against Christianity should be met, and every argument for it be exhibited, and all the graces and sublimities of its doctrine be held out to their merited admiration. And if it be true, as it certainly is, that throughout the whole of this process, a man may be carried rejoicingly along from the mere indulgence of his taste, and the mere play and exercise of his understanding; while conscience is untouched, and the supremacy of moral claims upon the heart and the conduct is practically disowned by him—it is further right that this should be adverted to; and that such a melancholy unhingement in the constitution of man should be fully laid open; and that he should be driven out of the seductive complacency which he is so apt to cherish, merely because he delights in the loveliness of the song; and that he should be urged with the imperiousness of a demand which still remains unsatisfied, to turn him from the corrupt indifference of nature, and to become personally a religious man; and that he should be assured how all the gratification he felt in listening to the word which respected the kingdom of God, will be of no avail, unless that kingdom come to himself in power—that it will only go to heighten the perversity of his character—that it will not extenuate his real and practical ungodliness, but will serve most fearfully to aggravate its condemnation.

With a religion so argumentable as ours, it may

be easy to gather out of it a feast for the human understanding. With a religion so magnificent as ours, it may be easy to gather out of it a feast for the human imagination. But with a religion so humbling, and so strict, and so spiritual, it is not easy to mortify the pride, or to quell the strong enmity of nature; or to arrest the currency of the affections; or to turn the constitutional habits; or to pour a new complexion over the moral history; or to stem the domineering influence of things seen and things sensible; or to invest faith with a practical supremacy; or to give its objects such a vivacity of influence as shall overpower the near and the hourly impressions, that are ever emanating upon man from a seducing world. It is here that man feels himself treading upon the limit of his helplessness. It is here that he sees where the strength of nature ends; and the power of grace must either be put forth, or leave him to grope his darkling way without one inch of progress towards the life and the substance of Christianity. It is here that a barrier rises on the contemplation of the inquirer—the barrier of separation between the carnal and the spiritual, and on which he may idly waste the every energy which belongs to him in the enterprise of surmounting it. It is here, that after having walked the round of nature's acquisitions, and lavished upon the truth all his ingenuities, and surveyed it in its every palpable character of grace and majesty, he will still feel himself on a level with the simplest and most untutored of the species. He needs the power of a living manifestation. He needs the anointing which remaineth. He needs that which fixes and perpetuates a stable revolution upon the character, and in virtue of which he may be advanced from the state of one who hears and is delighted, to the state of one who hears and is a doer. How strikingly is the experience even of vigorous and accomplished nature at one on this point with the announcements of revelation, that to work this change, there must be the putting forth of a peculiar agency; and that it is an agency, which, withheld from the exercise of loftiest talent, is often brought down on an impressed audience, through the humblest of all instrumentality, with the demonstration of the Spirit and with power.

Think it not enough, that you carry in your bosom an expanding sense of the magnificence of creation. But pray for a subduing sense of the authority of the Creator. Think it not enough, that with the justness of a philosophical discernment, you have traced that boundary which hems in all the possibilities of human attainment, and have found that all beyond it is a dark and fathomless unknown. But let this modesty of science be carried, as in consistency it ought, to the question of revelation, and let all the antipathies of nature be schooled to acquiescence in the authentic testimonies of the Bible. Think it not enough, that you have looked with sensibility and wonder at the representation of God throned in immensity, yet combining, with the vastness of his entire superintendance, a most thorough inspection into all the minute and countless diversities of existence. Think of your own heart as one of these diversities; and that he ponders all its tendencies; and has an eye upon all its movements; and marks all its waywardness; and, God of judgment as he is, records its every secret, and its every sin, in the book of his remembrance. Think it not enough, that you have been led to associate a grandeur with the salvation of the New Testament, when made to understand that it draws upon it the regards of an arrested universe. How is it arresting your own mind? What has been the earnestness of your personal regards towards it? And tell us, if all its faith, and all its repentance, and all its holiness, are not disowned by you? Think it not enough, that you have felt a sentimental charm when angels were pictured to your fancy as beckoning you to their mansions, and anxiously looking to the every symptom of your grace and reformation. Be constrained by the power of all this tenderness, and yield yourselves up in a practical obedience to the call of the Lord God, merciful and gracious. Think it not enough, that you have shared for a moment in the deep and busy interest of that arduous conflict which is now going on for a moral ascendancy over the species. Remember that the conflict is for each of you individually; and let this alarm you into a watchfulness against the power of every temptation, and a cleaving dependence upon Him through whom alone you will be more than conquerors. Above all, forget not, that while you only hear and are delighted, you are still under nature's powerlessness and nature's condemnation —and that the foundation is not laid, the mighty and essential change is not accomplished, the transition from death unto life is not undergone, the saving faith is not formed, nor the passage taken from darkness to the marvellous light of the gospel, till you are both hearers of the word and doers also. "For if any be a hearer of the word, and not a doer, he is like unto a man beholding his natural face in a glass: for he beholdeth himself, and goeth his way, and straightway forgetteth what manner of man he was."

APPENDIX.

The writer of these Discourses has drawn up the following compilation of passages from Scripture, as serving to illustrate or to confirm the leading arguments which have been employed in each separate division of his subject.

DISCOURSE I.

In the beginning God created the heaven and the earth.—Gen. i. 1.

Thus the heavens and the earth were finished, and all the host of them.—Gen. ii. 1.

Behold, the heaven, and the heaven of heavens, is the Lord's thy God, the earth also, with all that therein is.—Deut. x. 14.

There is none like unto the God of Jeshurun, who rideth upon the heaven in thy help, and in his excellency on the sky.—Deut. xxxiii. 26.

And Hezekiah prayed before the Lord, and said, O Lord God of Israel, which dwellest between the cherubims, thou art the God, even thou alone, of all the kingdoms of the earth; thou hast made heaven and earth.—2 Kings xix. 15.

For all the gods of the people are idols: but the Lord made the heavens.—1 Chron. xvi. 26.

Thou, even thou, art Lord alone: thou hast

made heaven, the heaven of heavens, with all their host, the earth, and all things that are therein, the seas, and all that is therein; and thou preservest them all; and the host of heaven worshippeth thee.—Nehemiah ix. 6.

Which alone spreadeth out the heavens, and treadeth upon the waves of the sea; which maketh Arcturus, Orion, and Pleiades, and the chambers of the south.—Job ix. 8, 9.

He stretcheth out the north over the empty place, and hangeth the earth upon nothing.—Job xxvi. 7.

By his Spirit he hath garnished the heavens.—Job xxvi. 13.

The heavens declare the glory of God; and the firmament showeth his handy-work.—Psalm xix. 1.

By the word of the Lord were the heavens made; and all the host of them by the breath of his mouth. —Psalm xxxiii. 6.

Of old hast thou laid the foundation of the earth; and the heavens are the work of thy hands.—Psalm cii. 25.

Who coverest thyself with light as with a garment; who stretchest out the heavens like a curtain. —Psalm civ. 2.

He appointed the moon for seasons; the sun knoweth his going down.—Psalm civ. 19.

Ye are blessed of the Lord, which made heaven and earth. The heaven, even the heavens, are the Lord's: but the earth hath he given to the children of men.—Psalm cxv. 15, 16.

My help cometh from the Lord, which made heaven and earth.—Psalm cxxi. 2.

Our help is in the name of the Lord, who made heaven and earth.—Psalm cxxiv. 8.

The Lord, that made heaven and earth, bless thee out of Zion.—Psalm cxxxiv. 3.

Which made heaven, and earth, the sea, and all that therein is.—Psalm cxlvi. 6.

The Lord by wisdom hath founded the earth; by understanding hath he established the heavens. —Prov. iii. 19.

Who hath measured the waters in the hollow of his hand, and meted out heaven with the span, and comprehended the dust of the earth in a measure, and weighed the mountains in scales, and the hills in a balance?—Isa. xl. 12.

It is he that sitteth upon the circle of the earth, and the inhabitants thereof are as grasshoppers; that stretcheth out the heavens as a curtain, and spreadeth them out as a tent to dwell in.—Isa. xl. 22.

Thus saith God the Lord, he that created the heavens, and stretched them out; he that spread forth the earth, and that which cometh out of it; he that giveth breath unto the people upon it, and spirit to them that walk therein.—Isa. xlii. 5.

Thus saith the Lord, thy Redeemer, and he that formed thee from the womb, I am the Lord that maketh all things; that stretcheth forth the heavens alone; that spreadeth abroad the earth by himself.—Isa. xliv. 24.

I have made the earth, and created man upon it: I, even my hands, have stretched out the heavens, and all their host have I commanded.—Isa. xlv. 12.

For thus saith the Lord that created the heavens, God himself that formed the earth, and made it; he hath established it, he created it not in vain, he formed it to be inhabited.—Isa. xlv. 18.

Mine hand also hath laid the foundation of the earth, and my right hand hath spanned the heavens: when I call unto them, they stand up together.—Isa. xlviii. 13.

He hath made the earth by his power, he hath established the world by his wisdom, and hath stretched out the heavens by his discretion.—Jer. x. 12.

Ah Lord God! behold, thou hast made the heaven and the earth by thy great power and stretched-out arm, and there is nothing too hard for thee.—Jer. xxxii. 17.

He hath made the earth by his power, he hath established the world by his wisdom, and hath stretched out the heaven by his understanding.—Jer. li. 15.

It is he that buildeth his stories in the heaven, and hath founded his troop in the earth; he that calleth for the waters of the sea, and poureth them out upon the face of the earth: The Lord is his name.—Amos ix. 6.

We also are men of like passions with you, and preach unto you, that ye should turn from these vanities unto the living God, which made heaven, and earth, and the sea, and all things that are therein.—Acts xiv. 15.

Hath in these last days spoken unto us by his Son, whom he hath appointed heir of all things, by whom also he made the worlds.—Heb. i. 2.

Thou, Lord, in the beginning hast laid the foundation of the earth; and the heavens are the works of thine hands.—Heb. i. 10.

Through faith we understand that the worlds were framed by the word of God.—Heb. xi. 3.

DISCOURSE II.

The secret things belong unto the Lord our God; but those things which are revealed belong unto us and to our children for ever, that we may do all the words of this law.—Deut. xxix. 29.

I would seek unto God, and unto God would I commit my cause; which doeth great things and unsearchable; marvellous things without number.—Job v. 8, 9.

Which doeth great things past finding out; yea, and wonders without number.—Job ix. 10.

Canst thou by searching find out God? canst thou find out the Almighty unto perfection?—Job xi. 7.

Hast thou heard the secret of God? and dost thou restrain wisdom to thyself?—Job xv. 8.

Lo, these are parts of his ways; but how little a portion is heard of him? but the thunder of his power who can understand?—Job xxvi. 14.

Behold, God is great, and we know him not, neither can the number of his years be searched out.—Job xxxvi. 26.

God thundereth marvellously with his voice;

great things doeth he, which we cannot comprehend. —Job xxxvii. 5.

Touching the Almighty, we cannot find him out: he is excellent in power, and in judgment, and in plenty of justice.—Job xxxvii. 23.

Thy way is in the sea, and thy path in the great waters, and thy footsteps are not known.—Psalm lxxvii. 19.

Great is the Lord, and greatly to be praised; and his greatness is unsearchable.—Psalm cxlv. 3.

For my thoughts are not your thoughts, neither are your ways my ways, saith the Lord. For as the heavens are higher than the earth, so are my ways higher than your ways, and my thoughts than your thoughts.—Isa. lv. 8, 9.

Verily I say unto you, Except ye be converted, and become as little children, ye shall not enter into the kingdom of heaven.—Matt. xviii. 3.

Verily I say unto you, Whosoever shall not receive the kingdom of God as a little child shall in nowise enter therein.—Luke xviii. 17.

O the depth of the riches both of the wisdom and knowledge of God! how unsearchable are his judgments, and his ways past finding out! For who hath known the mind of the Lord? or who hath been his counsellor?—Rom. xi. 33, 34.

Let no man deceive himself. If any man among you seemeth to be wise in this world, let him become a fool, that he may be wise.—1 Cor. iii. 18.

For if a man thinketh himself to be something, when he is nothing, he deceiveth himself.—Gal vi. 3.

Beware lest any man spoil you through philosophy and vain deceit, after the tradition of men, after the rudiments of the world, and not after Christ.—Col. ii. 8.

O Timothy, keep that which is committed to thy trust, avoiding profane and vain babblings, and oppositions of science falsely so called.—1 Tim. vi. 20

DISCOURSE III.

But will God indeed dwell on the earth? Behold, the heaven, and heaven of heavens, cannot contain thee; how much less this house that I have builded! Yet have thou respect unto the prayer of thy servant, and to his supplication, O Lord my God, to hearken unto the cry and to the prayer which thy servant prayeth before thee to-day: that thine eyes may be open toward this house night and day, even toward the place of which thou hast said, My name shall be there; that thou mayest hearken unto the prayer which thy servant shall make toward this place.—1 Kings viii. 27, 28, 29.

For he looketh to the ends of the earth, and seeth under the whole heaven.—Job xxviii. 24.

For his eyes are upon the ways of man, and he seeth all his goings.—Job xxxiv. 21.

Though the Lord be high, yet hath he respect unto the lowly.—Psalm cxxxviii. 6.

O Lord, thou hast searched me, and known me. Thou knowest my down-sitting and mine up-rising; thou understandest my thought afar off. Thou compassest my path, and my lying down, and art acquainted with all my ways. For there is not a word in my tongue, but, lo, O Lord, thou knowest it altogether. Thou hast beset me behind and before, and laid thine hand upon me. Such knowledge is too wonderful for me; it is high, I cannot attain unto it. Whither shall I go from thy Spirit? or whither shall I flee from thy presence?—Psalm cxxxix. 1—7.

How precious also are thy thoughts unto me, O God! how great is the sum of them! If I should count them, they are more in number than the sand: when I awake, I am still with thee.—Psalm cxxxix. 17, 18.

The eyes of the Lord are in every place, beholding the evil and the good.—Prov. xv. 3.

Can any hide himself in secret places that I shall not see him? saith the Lord: do not I fill heaven and earth? saith the Lord.—Jer. xxiii. 24.

Behold the fowls of the air: for they sow not, neither do they reap, nor gather into barns; yet your heavenly Father feedeth them. Are ye not much better than they? And why take ye thought for raiment? Consider the lilies of the field, how they grow; they toil not, neither do they spin: and yet I say unto you, That even Solomon, in all his glory, was not arrayed like one of these. Wherefore, if God so clothe the grass of the field, which to-day is, and to-morrow is cast into the oven, shall he not much more clothe you, O ye of little faith?—Matt. vi. 26, 28, 29, 30.

But the very hairs of your head are all numbered. —Matt. x. 30.

Neither is there any creature that is not manifest

in his sight: but all things are naked and opened unto the eyes of him with whom we have to do.—Heb. iv. 13.

DISCOURSE IV.

And he dreamed, and behold a ladder set up on the earth, and the top of it reached to heaven; and behold the angels of God ascending and descending on it.—Gen. xxviii. 12.

For a thousand years in thy sight are but as yesterday when it is past, and as a watch in the night.—Psalm xc. 4.

Lift up your eyes to the heavens, and look upon the earth beneath; for the heavens shall vanish away like smoke, and the earth shall wax old like a garment, and they that dwell therein shall die in like manner: but my salvation shall be for ever, and my righteousness shall not be abolished.—Isa. li. 6.

For the Son of man shall come in the glory of his Father, with his angels; and then he shall reward every man according to his works.—Matt. xvi. 27.

When the Son of man shall come in his glory, and all the holy angels with him, then shall he sit upon the throne of his glory.—Matt. xxv. 31.

Also I say unto you, Whosoever shall confess me before men, him shall the Son of man also confess before the angels of God: but he that denieth me before men, shall be denied before the angels of God.—Luke xii. 8, 9.

And he saith unto him, Verily, verily, I say

unto you, Hereafter ye shall see heaven open, and the angels of God ascending and descending upon the Son of man.—John i. 51.

We are made a spectacle unto the world, and to angels, and to men.—1 Cor. iv. 9.

Wherefore God also hath highly exalted him, and given him a name which is above every name; that at the name of Jesus every knee should bow, of things in heaven, and things in earth, and things under the earth; and that every tongue should confess that Jesus Christ is Lord, to the glory of God the Father.—Phil. ii. 9, 10, 11.

When the Lord Jesus shall be revealed from heaven with his mighty angels.—2 Thess. i. 7.

And, without controversy, great is the mystery of godliness: God was manifest in the flesh, justified in the Spirit, seen of angels, preached unto the Gentiles, believed on in the world, received up into glory.—1 Tim. iii. 16.

I charge thee before God, and the Lord Jesus Christ, and the elect angels, that thou observe these things.—1 Tim. v. 21.

And again, when he bringeth in the first-begotten into the world, he saith, And let all the angels of God worship him.—Heb. i. 6.

But ye are come unto Mount Zion, and unto the city of the living God, the heavenly Jerusalem, and to an innumerable company of angels, to the general assembly and church of the firstborn, which are written in heaven, and to God the Judge of all, and to the spirits of just men made perfect, and to Jesus the mediator of the new covenant.—Heb. xii. 22, 23, 24

But, beloved, be not ignorant of this one thing, that one day is with the Lord as a thousand years, and a thousand years as one day. The Lord is not slack concerning his promise, as some men count slackness; but is long-suffering to us-ward, not willing that any should perish, but that all should come to repentance. But the day of the Lord will come as a thief in the night; in the which the heavens shall pass away with a great noise, and the elements shall melt with fervent heat, the earth also, and the works that are therein, shall be burnt up.—2 Peter iii. 8, 9, 10.

And the angel which I saw stand upon the sea and upon the earth lifted up his hand to heaven, and sware by him that liveth for ever and ever, who created heaven, and the things that therein are, and the earth, and the things that therein are, and the sea, and the things which are therein, that there should be time no longer.—Rev. x. 5, 6.

And the third angel followed them, saying with a loud voice, If any man worship the beast and his image, and receive his mark in his forehead, or in his hand, the same shall drink of the wine of the wrath of God, which is poured out without mixture into the cup of his indignation; and he shall be tormented with fire and brimstone in the presence of the holy angels, and in the presence of the Lamb. —Rev. xiv. 9, 10.

And I saw a great white throne, and him that sat on it, from whose face the earth and the heaven fled away; and there was found no place for them. —Rev. xx. 11.

DISCOURSE V.

And Nathan departed unto his house: and the Lord struck the child that Uriah's wife bare unto David, and it was very sick. David therefore besought God for the child; and David fasted, and went in, and lay all night upon the earth. And the elders of his house arose, and went to him, to raise him up from the earth: but he would not, neither did he eat bread with them. And it came to pass on the seventh day, that the child died. And the servants of David feared to tell him that the child was dead; for they said, Behold, while the child was yet alive, we spake unto him, and he would not hearken unto our voice: how will he then vex himself, if we tell him that the child is dead? But when David saw that his servants whispered, David perceived that the child was dead: therefore David said unto his servants, Is the child dead? And they said, He is dead. Then David arose from the earth, and washed, and anointed himself, and changed his apparel, and came into the house of the Lord, and worshipped: then he came to his own house; and when he required, they set bread before him, and he did eat. Then said his servants unto him, What thing is this that thou hast done? Thou didst fast and weep for the child, while it was alive; but when the child was dead, thou didst rise and eat bread. And he said, While the child was yet alive, I fasted and wept: for I said, Who can tell whether God will be gracious to me, that the child may live? But now he is dead, wherefore should I fast? can I bring him back again? I shall go to him, but he shall not return to me.—2 Sam. xii. 15—23.

The angel of the Lord encampeth round about them that fear him, and delivereth them.—Psalm xxxiv. 7.

For he shall give his angels charge over thee, to keep thee in all thy ways.—Psalm xci. 11.

And he shall send his angels with a great sound of a trumpet, and they shall gather together his elect from the four winds, from one end of heaven to the other.—Matt. xxiv. 31.

Likewise, I say unto you, There is joy in the presence of the angels of God over one sinner that repenteth.—Luke xv. 10.

Are they not all ministering spirits, sent forth to minister for them who shall be heirs of salvation? —Heb. i. 14.

DISCOURSE VI.

Then was Jesus led up of the Spirit into the wilderness, to be tempted of the devil.—Matt. iv. 1.

The enemy that sowed them is the devil; the harvest is the end of the world; and the reapers are the angels. The Son of man shall send forth his angels, and they shall gather out of his kingdom all things that offend, and them which do iniquity. —Matt. xiii. 39, 41.

Then shall he say also unto them on the left hand, Depart from me, ye cursed, into everlasting fire, prepared for the devil and his angels.—Matt. xxv. 41.

And in the synagogue there was a man which had a spirit of an unclean devil, and cried out with a loud voice, saying, Let us alone; what have we to do with thee, thou Jesus of Nazareth? art thou come to destroy us? I know thee who thou art: the Holy One of God.—Luke iv. 33, 34.

Those by the way-side are they that hear; then cometh the devil, and taketh away the word out of their hearts, lest they should believe and be saved. —Luke viii. 12.

But he, knowing their thoughts, said unto them, Every kingdom divided against itself is brought to desolation; and a house divided against a house falleth. If Satan also be divided against himself, how shall his kingdom stand? because ye say that I cast out devils through Beelzebub.—Luke xi. 17, 18.

Ye are of your father the devil, and the lusts of your father ye will do: he was a murderer from the beginning, and abode not in the truth, because there is no truth in him. When he speaketh a lie, he speaketh of his own: for he is a liar, and the father of it.—John viii. 44.

And supper being ended, (the devil having now put into the heart of Judas Iscariot, Simon's son, to betray him.)—John xiii. 2.

But Peter said, Ananias, why hath Satan filled thine heart to lie to the Holy Ghost, and to keep back part of the price of the land?—Acts v. 3.

To open their eyes, and to turn them from darkness to light, and from the power of Satan unto God, that they may receive forgiveness of sins, and inheritance among them which are sanctified by faith that is in me.—Acts xxvi. 18.

And the God of peace shall bruise Satan under your feet shortly. The grace of our Lord Jesus Christ be with you. Amen.—Rom. xvi. 20.

Lest Satan should get an advantage of us: for we are not ignorant of his devices.—2 Cor. ii. 11.

In whom the god of this world hath blinded the minds of them which believe not, lest the light of the glorious gospel of Christ, who is the image of God, should shine unto them.—2 Cor. iv. 4.

Wherein in time past ye walked according to the course of this world, according to the prince of the power of the air, the spirit that now worketh in the children of disobedience.—Eph. ii. 2.

Put on the whole armour of God, that ye may be able to stand against the wiles of the devil. For we wrestle not against flesh and blood, but against principalities, against powers, against the rulers of the darkness of this world, against spiritual wickedness in high places.—Eph. vi. 11, 12.

For some are already turned aside after Satan.—1 Tim. v. 15.

Forasmuch then as the children are partakers of flesh and blood, he also himself likewise took part of the same; that through death he might destroy him that had the power of death, that is, the devil. —Heb. ii. 14.

Submit yourselves therefore to God. Resist the devil, and he will flee from you.—James iv. 7.

Be sober, be vigilant; because your adversary the devil, as a roaring lion, walketh about, seeking whom he may devour: whom resist steadfast in the faith, knowing that the same afflictions are accomplished in your brethren that are in the world.—1 Pet. v. 8, 9.

He that committeth sin is of the devil; for the devil sinneth from the beginning. For this purpose the Son of God was manifested, that he might destroy the works of the devil.—In this the children of God are manifest, and the children of the devil: whosoever doeth not righteousness is not of God, neither he that loveth not his brother —1 John iii. 8, 10.

Ye are of God, little children, and have overcome them; because greater is he that is in you, than he that is in the world.—1 John iv. 4.

And the angels which kept not their first estate, but left their own habitation, he hath reserved in everlasting chains, under darkness, unto the judgment of the great day.—Jude 6.

He that overcometh, the same shall be clothed in white raiment; and I will not blot out his name out of the book of life, but I will confess his name before my Father, and before his angels.—Rev. iii. 5.

And there was war in heaven: Michael and his angels fought against the dragon; and the dragon fought and his angels, and prevailed not; neither was their place found any more in heaven. And the great dragon was cast out, that old serpent, called the Devil, and Satan, which deceiveth the whole world; he was cast out into the earth, and his angels were cast out with him. Therefore rejoice, ye heavens, and ye that dwell in them. Woe to the inhabiters of the earth and of the sea! for the devil is come down unto you, having great wrath, because he knoweth that he hath but a short time.—Rev. xii. 7, 8, 9, 12.

And he laid hold on the dragon, that old serpent, which is the Devil, and Satan, and bound him a thousand years. And when the thousand years are expired, Satan shall be loosed out of his prison. And the devil that deceived them was cast into the lake of fire and brimstone, where the beast and the false prophet are, and shall be tormented day and night for ever and ever.—Rev. xx. 2, 7, 10.

DISCOURSE VII.

Therefore, whosoever heareth these sayings of mine, and doeth them, I will liken him unto a wise man, which built his house upon a rock; and the rain descended, and the floods came, and the winds blew, and beat upon that house; and it fell not: for it was founded upon a rock. And every one that heareth these sayings of mine, and doeth them not, shall be likened unto a foolish man, which built his house upon the sand: and the rain descended, and the floods came, and the winds blew: and beat upon that house; and it fell: and great was the fall of it.—Matt. vii. 24—27.

At that time Jesus answered and said, I thank thee, O Father, Lord of heaven and earth, because thou hast hid these things from the wise and prudent, and hast revealed them unto babes.—Matt. xi. 25.

Then shall ye begin to say, We have eaten and

drunk in thy presence, and thou hast taught in our streets. But he shall say, I tell you, I know you not whence ye are: depart from me, all ye workers of iniquity.—Luke xiii. 26, 27.

For not the hearers of the law are just before God, but the doers of the law shall be justified.—Rom. ii. 13.

And I, brethren, when I came to you, came not with excellency of speech, or of wisdom, declaring unto you the testimony of God: for I determined not to know any thing among you, save Jesus Christ, and him crucified. And my speech and my preaching was not with enticing words of man's wisdom, but in demonstration of the Spirit and of power; that your faith should not stand in the wisdom of men, but in the power of God. Now we have received, not the spirit of the world, but the Spirit which is of God; that we might know the things that are freely given to us of God. Which things also we speak, not in the words which man's wisdom teacheth, but which the Holy Ghost teacheth; comparing spiritual things with spiritual. But the natural man receiveth not the things of the Spirit of God; for they are foolishness unto him: neither can he know them, because they are spiritually discerned.—1 Cor. ii. 1, 2, 4, 5, 12, 13, 14.

For the wisdom of this world is foolishness with God.—1 Cor. iii. 19.

For the kingdom of God is not in word, but in power.—1 Cor. iv. 20.

Forasmuch as ye are manifestly declared to be the epistle of Christ, ministered by us, written not with ink, but with the Spirit of the living God; not in tables of stone, but in fleshly tables of the heart. Not that we are sufficient of ourselves to think any thing as of ourselves; but our sufficiency is of God; who also hath made us able ministers of the New Testament; not of the letter, but of the spirit: for the letter killeth, but the spirit giveth life.—2 Cor. iii. 3, 5, 6.

That the God of our Lord Jesus Christ, the Father of glory, may give unto you the spirit of wisdom and revelation in the knowledge of him: the eyes of your understanding being enlightened; that ye may know what is the hope of his calling, and what the riches of the glory of his inheritance in the saints, and what is the exceeding greatness of his power to us-ward who believe, according to the working of his mighty power.—Eph. i. 17, 18, 19.

And you hath he quickened, who were dead in trespasses and sins. For we are his workmanship, created in Christ Jesus unto good works.—Eph. ii. 1, 10.

For our gospel came not unto you in word only, but also in power, and in the Holy Ghost, and in much assurance.—1 Thess. i. 5.

Of his own will begat he us with the word of truth, that we should be a kind of first-fruits of his creatures. But be ye doers of the word, and not hearers only, deceiving your ownselves. For if any be a hearer of the word, and not a doer, he is like unto a man beholding his natural face in a glass: for he beholdeth himself, and goeth his way, and straightway forgetteth what manner of man he was. But whoso looketh into the perfect law of liberty, and continueth therein, he being not a forgetful hearer, but a doer of the work, this man shall be blessed in his deed.—James i. 18, 22, 25.

But ye are a chosen generation, a royal priesthood, an holy nation, a peculiar people; that ye should show forth the praises of him who hath called you out of darkness into his marvellous light.—1 Peter ii. 9.

But ye have an unction from the Holy One, and ye know all things. But the anointing which ye have received of him abideth in you; and ye need not that any man teach you: but as the same anointing teacheth you of all things, and is truth, and is no lie, and even as it hath taught you, ye shall abide in him.—1 John ii. 20, 27.

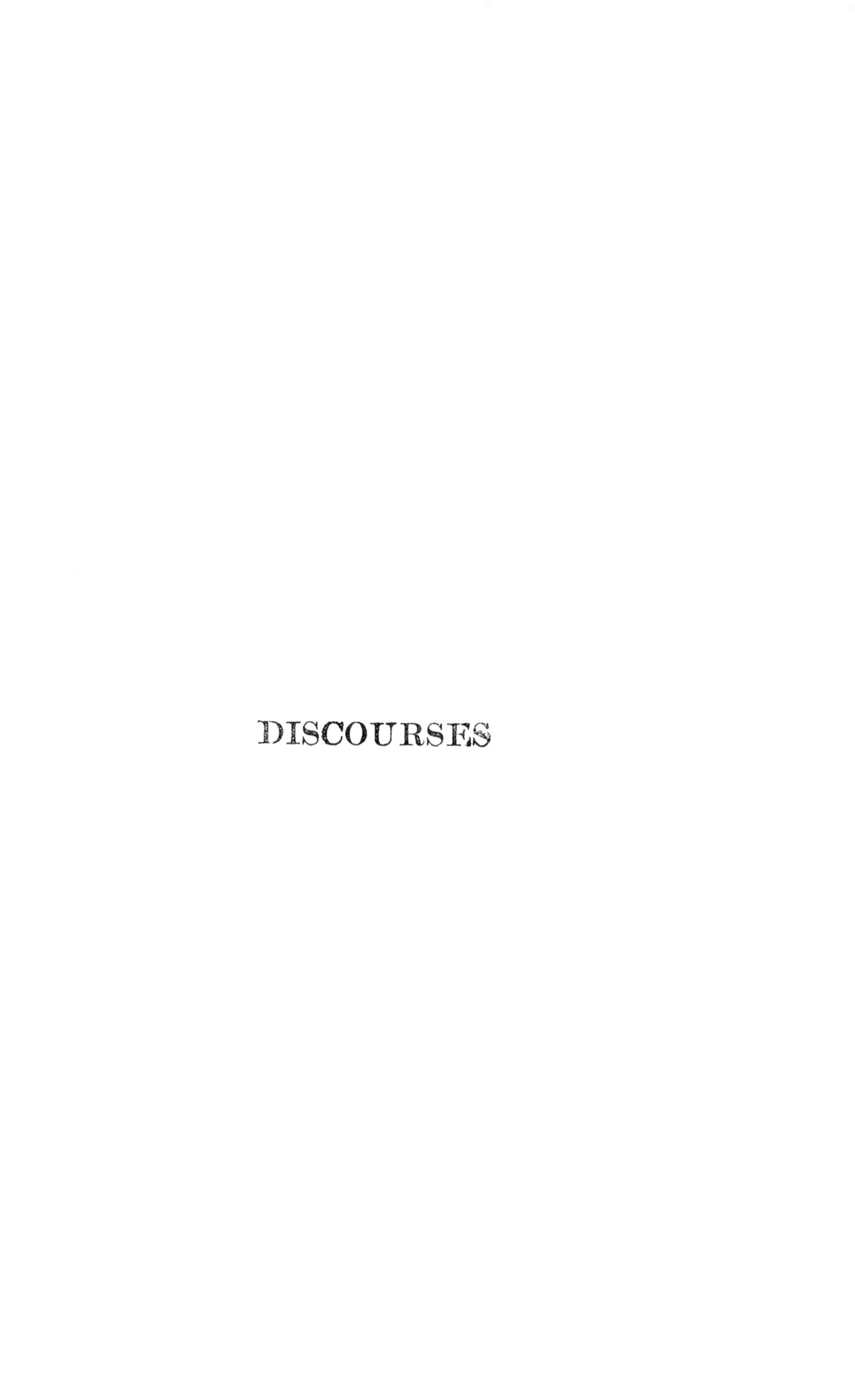

DISCOURSES

DISCOURSE I.

THE CONSTANCY OF GOD IN HIS WORKS AN ARGUMENT FOR THE FAITHFULNESS OF GOD IN HIS WORD.

"For ever, O Lord, thy word is settled in heaven. Thy faithfulness is unto all generations: thou hast established the earth, and it abideth. They continue this day according to thine ordinances: for all are thy servants."—PSALM cxix. 89, 90, 91.

IN these verses there is affirmed to be an analogy between the word of God and the works of God. It is said of His word, that it is settled in heaven, and that it sustains its faithfulness from one generation to another. It is said of His works, and more especially of those that are immediately around us, even of the earth which we inhabit, that as it was established at the first so it abideth afterwards. And then, as if to perfect the assimilation between them, it is said of both in the 91st verse, "They continue this day according to thine ordinances, for all are thy servants;" thereby identifying the sureness of that word which proceeded from His lips, with the unfailing constancy of that Nature which was formed and is upholden by His hands.

The constancy of Nature is taught by universal experience, and even strikes the popular eye as the most characteristic of those features which have been impressed upon her. It may need the aid of philosophy to learn how unvarying Nature is in all her processes—how even her seeming anomalies can be traced to a law that is inflexible—how what might appear at first to be the caprices of her way-wardness, are, in fact, the evolutions of a mechanism that never changes—and that the more thoroughly she is sifted and put to the test by the interrogations of the curious, the more certainly will they find that she walks by a rule which knows no abatement, and perseveres with obedient footstep in that even course, from which the eye of strictest scrutiny, has never yet detected one hair-breadth of deviation. It is no longer doubted by men of science, that every remaining semblance of irregularity in the universe is due, not to the fickleness of Nature, but to the ignorance of man—that her most hidden movements are conducted with a uniformity as rigorous as Fate—that even the fitful agitations of the weather have their law and their principle—that the intensity of every breeze, and the number of drops in every shower, and the formation of every cloud, and all the occurring alternations of storm and sunshine, and the endless shiftings of tempera-ture, and those tremulous varieties of the air which our instruments have enabled us to discover but have not enabled us to explain—that still, they follow each other by a method of succession, which, though greatly more intricate, is yet as absolute in itself as the order of the seasons, or the mathemati-cal courses of astronomy. This is the impression of every philosophical mind with regard to Nature, and it is strengthened by each new accession that is made to science. The more we are acquainted with her, the more are we led to recognise her constancy; and to view her as a mighty though complicated machine, all whose results are sure, and all whose workings are invariable.

But there is enough of patent and palpable regularity in Nature, to give also to the popular mind, the same impression of her constancy. There is a gross and general experience that teaches the same lesson, and that has lodged in every bosom a kind of secure and steadfast confidence in the uniformity of her processes. The very child knows and proceeds upon it. He is aware of an abiding character and property in the elements around him —and has already learned as much of the fire, and the water, and the food that he eats, and the firm ground that he treads upon, and even of the gravitation by which he must regulate his postures and his movements, as to prove, that, infant though he be, he is fully initiated in the doctrine, that Nature has her laws and her ordinances, and that she continueth therein. And the proofs of this are ever multiplying along the journey of human observation: insomuch, that when we come to manhood, we read of Nature's constancy throughout every department of the visible world. It meets us wherever we turn our eyes. Both the day and the night bear witness to it. The silent revolutions of the firmament give it their pure testimony. Even those appearances in the heavens, at which superstition stood aghast, and imagined that Nature was on the eve of giving way, are the proudest trophies of that stability which reigns throughout her processes— of that unswerving consistency wherewith she prosecutes all her movements. And the lesson that is thus held forth to us from the heavens above, is responded to by the earth below; just as the tides of ocean wait the footsteps of the moon, and, by an attendance kept up without change or intermission for thousands of years, would seem to connect the regularity of earth with the regularity of heaven. But, apart from these greater and simpler energies, we see a course and a uniformity everywhere. We recognise it in the mysteries of vegetation. We follow it through the successive stages of growth, and maturity, and decay, both in plants and animals. We discern it still more palpably in that beautiful circulation of the element of water, as it rolls its way by many thousand channels to the ocean—and, from the surface of this expanded reservoir, is again uplifted to the higher regions of the atmosphere—and is there dispersed in light and fleecy magazines over the four quarters of the globe—and at length accomplishes its orbit, by falling in showers on a world that waits to be refreshed by it. And all goes to impress us with the regularity of Nature, which in fact teems, throughout all its varieties, with power, and principle, and uniform laws of operation—and is viewed by us as a vast laboratory, all the progressions of which have a rigid and unfailing necessity stamped upon them.

Now, this contemplation has at times served to foster the atheism of philosophers. It has led them to deify Nature, and to make her immutability stand in the place of God. They seem imprest with the imagination, that had the Supreme Cause been a Being who thinks, and wills, and acts as man does, on the impulse of a felt and a present motive, there would be more the appearance of spontaneous activity, and less of mute and unconscious mechanism in the administrations of the universe. It is the very unchangeableness of Nature, and the steadfastness of those great and mighty processes wherewith no living power that is superior to Nature, and is able to shift or to control her, is seen to interfere—it is this which seems to have imprest the notion of some blind and eternal fatality on certain men of loftiest but deluded genius. And, accordingly, in France, where the physical sciences have, of late, been the most cultivated, have there also been the most daring avowals of atheism. The universe has been affirmed to be an everlasting and indestructible effect; and from the abiding constancy that is seen in Nature, through all her departments, have they inferred, that thus it has always been, and that thus it will ever be.

But this atheistical impression that is derived from the constancy of Nature is not peculiar to the disciples of philosophy. It is the familiar and the practical impression of every-day life. The world is apprehended to move on steady and unvarying principles of its own; and these secondary causes have usurped, in man's estimation, the throne of the Divinity. Nature in fact is personified into God: and as we look to the performance of a machine without thinking of its maker,—so the very exactness and certainty, wherewith the machinery of creation performs its evolutions, has thrown a disguise over the agency of the Creator. Should God interpose by miracle, or interfere by some striking and special manifestation of providence, then man is awakened to the recognition of him.

But he loses sight of the Being who sits behind these visible elements, while he regards those attributes of constancy and power which appear in the elements themselves. They see no demonstration of a God, and they feel no need of Him, while such unchanging, and such unfailing energy continues to operate in the visible world around them; and we need not go to the schools of ratiocination in quest of this infidelity, but may detect it in the bosoms of simple and unlettered men, who, unknown to themselves, make a god of Nature, and just because of Nature's constancy; having no faith in the unseen Spirit who originated all and upholds all, and that, because all things continue as they were from the beginning of the Creation.

Such has been the perverse effect of Nature's constancy on the alienated mind of man: but let us now attend to the true interpretation of it. God has, in the first instance, put into our minds a disposition to count on the uniformity of Nature, insomuch that we universally look for a recurrence of the same event in the same circumstances. This is not merely the belief of experience, but the belief of instinct. It is antecedent to all the findings of observation, and may be exemplified in the earliest stages of childhood. The infant who makes a noise on the table with his hand, for the first time, anticipates a repetition of the noise from a repetition of the stroke, with as much confidence as he who has witnessed, for years together, the unvariableness wherewith these two terms of the succession have followed each other. Or, in other words, God, by putting this faith into every human creature, and making it a necessary part of his mental constitution, has taught him at all times to expect the like result in the like circumstances. He has thus virtually told him what is to happen, and what he has to look for in every given condition—and by its so happening accordingly, He just makes good the veracity of His own declaration. The man who leads me to expect that which he fails to accomplish, I would hold to be a deceiver. God has so framed the machinery of my perceptions, as that I am led irresistibly to expect, that everywhere events will follow each other in the very train in which I have ever been accustomed to observe them—and when God so sustains the uniformity of Nature, that in every instance it is rigidly so, He is just manifesting the faithfulness of his character. Were it otherwise, he would be practising a mockery on the expectation which He Himself had inspired. God may be said to have promised to every human being, that Nature will be constant—if not by the whisper of an inward voice to every heart, at least by the force of an uncontrollable bias which He has impressed on every constitution. So that, when we behold Nature keeping by its constancy, we behold the God of Nature keeping by His faithfulness—and the system of visible things, with its general laws, and its successions which are invariable, instead of an opaque materialism to intercept from the view of mortals the face of the Divinity, becomes the mirror which reflects upon them the truth that is unchangeable, the ordination that never fails.

Conceive that it had been otherwise—first, that man had no faith in the constancy of Nature—then how could all his experience have profited him? How could he have applied the recollections of his past, to the guidance of his future history? And, what would have been left to signalize the wisdom of mankind above that of veriest infancy? Or, suppose that he had the implicit faith in Nature's constancy, but that Nature was wanting in the fulfilment of it—that at every moment his intuitive reliance on this constancy, was met by some caprice or waywardness of Nature, which thwarted him in all his undertakings—that, instead of holding true to her announcements, she held the children of men in most distressful uncertainty, by the freaks and the falsities in which she ever indulged herself—and that every design of human foresight was thus liable to be broken up, by ever and anon the putting forth of some new fluctuation. Tell us, in this wild misrule of elements changing their properties, and events ever flitting from one method of succession to another, if man could subsist for a single day, when all the accomplishments without, were thus at war with all the hopes and calculations within. In such a chaos and conflict as this, would not the foundations of human wisdom be utterly subverted? Would not man, with his powerful and perpetual tendency to proceed on the constancy of Nature, be tempted, at all times, and by the very constitution of his being, to proceed upon a falsehood? It were the way, in fact, to turn the administration of Nature into a system of deceit. The lessons of to-day, would be falsified by the events of to-morrow. He were indeed the father of lies who would be the author of such a regimen as this

—and well may we rejoice in the strict order of the goodly universe which we inhabit, and regard it as a noble attestation to the wisdom and beneficence of its great Architect.

But it is more especially as an evidence of His truth, that the constancy of Nature is adverted to in our text. It is of His faithfulness unto all generations that mention is there made—and for the growth and the discipline of your piety, we know not a better practical habit than that of recognizing the unchangeable truth of God, throughout your daily and hourly experience of Nature's unchangeableness. Your faith in it is of His working—and what a condition would you have been reduced to, had the faith which is within, not been met by an entire and unexpected accordancy with the fulfilments that are without! He has not told you what to expect by the utterance of a voice—but He has taught you what to expect by the leadings and the intimations of a strong constitutional tendency—and, in virtue of this, there is not a human creature who does not believe, and almost as firmly as in his own existence, that fire will continue to burn, and water to cool, and matter to resist, and unsupported bodies to fall, and ocean to bear the adventurous vessel upon its surface, and the solid earth to uphold the tread of his footsteps; and that spring will appear again in her wonted smiles, and summer will glow into heat and brilliancy, and autumn will put on the same luxuriance as before, and winter, at its stated periods, revisit the world with her darkness and her storms. We cannot sum up these countless varieties of Nature;

but the firm expectation is, that throughout them all, as she has been established, so she will abide to the day of her final dissolution. And we call upon you to recognize in Nature's constancy, the answer of Nature's God to this expectation. All these material agents are, in fact, the organs by which He expresses His faithfulness to the world; and that unveering generality which reigns and continues everywhere, is but the perpetual demonstration of a truth that never varies, as well as of laws that never are rescinded. It is for us, that He upholds the world in all its regularity. It is for us, that He sustains so unviolably the march and the movement of those innumerable progressions, which are going on around us. It is in remembrance of His promises to us, that he meets all our anticipations of Nature's uniformity, with the evolutions of a law that is unalterable. It is because He is a God that cannot lie, that He will make no invasion on that wondrous correspondency which he himself hath instituted between the world that is without, and our little world of hopes, and projects, and anticipations that are within. By the constancy of Nature, He hath imprinted upon it the lesson of His own constancy—and that very characteristic wherewith some would fortify the ungodliness of their hearts, is the most impressive exhibition which can be given of God, as always faithful, and always the same.

This, then, is the real character which the constancy of Nature should lead us to assign to Him who is the Author of it. In every human understanding, He hath planted a universal instinct, by which all are led to believe, that Nature will persevere in her wonted courses, and that each succession of cause and effect which has been observed by us in the time that is past, will, while the world exists, be kept up invariably, and recur in the very same order through the time that is to come. This constancy, then, is as good as a promise that He has made unto all men, and all that is around us on earth or in heaven, proves how inflexibly the promise is adhered to. The chemist in his laboratory, as he questions Nature, may be almost said to put her to the torture, when tried in his hottest furnace, or probed by his searching analysis, to her innermost arcana, she by a spark or an explosion, or an effervescence, or an evolving substance, makes her distinct replies to his investigations. And he repeats her answer to all his fellows in philosophy, and they meet in academic state and judgment to reiterate the question, and in every quarter of the globe her answer is the same—so that, let the experiment, though a thousand times repeated, only be alike in all its circumstances, the result which cometh forth is as rigidly alike, without deficiency, and without deviation. We know how possible it is for these worshippers at the footstool of science, to make a divinity of matter; and that every new discovery of her secrets, should only rivet them more devotedly to her throne. But there is a God who liveth and sitteth there, and these unvarying responses of Nature, are all prompted by himself, and are but the utterances of His immutability. They are the replies of a God who never changes, and who hath adapted the whole materialism of creation to the constitution of every mind that He hath sent forth upon it. And to meet the expectation which He himself hath given of Nature's constancy, is He at each successive instant of time, vigilant and ready in every part of His vast dominions, to hold out to the eye of all observers, the perpetual and unfailing demonstration of it. The certainties of Nature and of Science, are in fact the vocables by which God announces His truth to the world—and when told how impossible it is, that nature can fluctuate, we are only told how impossible it is that the God of Nature can deceive us.

The doctrine that Nature is constant when thus related, as it ought to be, with the doctrine that God is true, might well strengthen our confidence in Him anew with every new experience of our history. There is not an hour or a moment, in which we may not verify the one—and, therefore, not an hour or a moment in which we may not invigorate the other. Every touch, and every look, and every taste, and every act of converse between our senses and the things that are without, brings home a new demonstration of the steadfastness of Nature, and along with it a new demonstration both of His steadfastness and of His faithfulness, who is the Governor of Nature. And the same lesson may be fetched from times and from places, that are far beyond the limits of our own personal history. It can be drawn from the retrospect of past ages, where, from the unvaried currency of those very processes which we now behold, we may learn the stability of all His ways, whose goings forth are of old, and from everlasting. It can be gathered from the most distant extremities of the earth, where Nature reigns with the same unwearied constancy, as it does around us—and where savages count as we do on a uniformity, from which she never falters. The lesson is commensurate with the whole system of things—and with an effulgence as broad as the face of creation, and as clear as the light which is poured over it, does it at once tell that Nature is unchangeably constant, and that God is unchangeably true.

And so it is, that in our text there are presented together, as if there was a tie of likeness between them—that the same God who is fixed as to the ordinances of Nature, is faithful as to the declarations of His word; and as all experience proves how firmly He may be trusted for the one, so is there an argument as strong as experience, to prove how firmly He may be trusted for the other. By his work in us, He hath awakened the expectation of a constancy in Nature, which He never disappoints. By His word to us, should He awaken the expectation of a certainty in His declarations, this he will never disappoint. It is because Nature is so fixed, that we apprehend the God of Nature to be so faithful. He who never falsifies the hope that hath arisen in every bosom, from the instinct which He Himself hath communicated, will never falsify the hope that shall arise in any bosom from the express utterance of His voice. Were He a God in whose hand the processes of Nature were ever shifting, then might we conceive Him a God from whose mouth the proclamations of grace had the like characters of variance and vacillation. But it is just because of our reliance on the one, that we feel so much of repose in our dependence upon the other—and the same God who is so unfailing in the ordinances of His creation, do we hold to be equally unfailing in the ordinances of His word.

And it is strikingly accordant with these views, that Nature never has been known to recede from her constancy, but for the purpose of giving place and demonstration to the authority of the word. Once, in a season of miracle, did the word take the precedency of Nature, but ever since hath Nature resumed her courses, and is now proving, by her steadfastness, the authority of that, which she then proved to be authentic by her deviations. When the word was first ushered in, Nature gave way for a period, after which she moves in her wonted order, till the present system of things shall pass away, and that faith which is now upholden by Nature's constancy, shall then receive its accomplishment at Nature's dissolution. And O how God magnifieth His word above all His name, when He tells that heaven and earth shall pass away, but that His word shall not pass away—and that while His creation shall become a wreck, not one jot or one tittle of His testimony shall fail. The world passeth away—but the word endureth for ever—and if the faithfulness of God stand forth so legibly on the face of the temporary world, how surely may we reckon on the faithfulness of that word, which has a vastly higher place in the counsels and fulfilments of eternity?

The argument may not be comprehended by all; but it will not be lost, should it lead any to feel a more emphatic certainty and meaning than before in the declarations of the Bible—and to conclude, that He, who for ages, hath stood so fixed to all His plans and purposes in Nature, will stand equally fixed to all that He proclaims, and to all that He promises in Revelation. To be in the hands of such a God, might well strike a terror into the hearts of the guilty—and that unrelenting death which, with all the sureness of an immutable law, is seen, before our eyes, to seize upon every individual of every species of our world, full well evinces how He, the uncompromising Lawgiver, will execute every utterance that He has made against the children of iniquity. And, on the other hand, how this very contemplation ought to encourage all who are looking to the announcements of the same God in the Gospel, and who perceive that there He has embarked the same truth, and the same unchangeableness, on the offers of mercy. All Nature gives testimony to this, that He cannot lie—and seeing that He has stamped such enduring properties on the elements even of our perishable world, never should I falter from that confidence which He hath taught me to feel, when I think of that property wherewith the blood which was shed for me, cleanseth from all sin; and of that property wherewith the body which was broken, beareth the burden of all its penalties. He who hath so nobly met the faith that He has given unto all in the constancy of Nature, by a uniformity which knows no abatement, will meet the faith that He has given unto any in the certainty of grace, by a fulfilment unto every believer, which knows no exception.

And it is well to remark the difference that there is between the explanation given in the text, of Nature's constancy, and the impression which the mere students or disciples of Nature have of it. It is because of her constancy that they have been led to invest her, as it were, in properties of her own; that they have given a kind of independent power and stability to matter; that in the various energies which lie scattered over the field of visible contemplation, they see a native inherent virtue, which never for a single moment is slackened or suspended—and therefore imagine, that as no force from without seems necessary to sustain, so as little, perhaps, is there need for any such force from without to originate. The mechanical certainty of all Nature's processes, as it appears in their eyes to supersede the demand for any upholding agency, so does it also supersede, in the silent imaginations of many, and according to the express and bold avowals of some, the demand for any creative agency. It is thus, that Nature is raised into a divinity, and has been made to reign over all, in the state and jurisdiction of an eternal fatalism; and proud Science, which by wisdom knoweth not God, hath, in her march of discovery, seized upon the invariable certainties of Nature, those highest characteristics of His authority and wisdom and truth, as the instruments by which to disprove and to dethrone him.

Now compare this interpretation of monstrous and melancholy atheism, with that which the Bible gives, why all things move so invariably. It is because that all are thy servants. It is because they are all under the bidding of a God who has purposes from which He never falters, and hath issued promises from which He never fails. It is because the arrangements of His vast and capacious household are already ordered for the best, and all the elements of Nature are the ministers by which He fulfils them. That is the master who has most honour and obedience from his domestics, throughout all whose ordinations, there runs a consistency from which He never deviates; and He best sustains His dignity in the midst of them, who, by mild but resistless sway, can regulate the successions of every hour, and affix His sure and appropriate service to every member of the family. It is when we see all, in any given time, at their respective places, and each distinct period of the day having its own distinct evolution of business or recreation, that we infer the wisdom of the instituted government, and how irrevocable the sanctions are by which it is upholden. The vexatious alternations of command and of countermand; the endless fancies of humour, and caprice, and waywardness, which ever and anon break forth, to the total overthrow of system; the perpetual innovations which none do foresee, and for which none, therefore, can possibly be prepared—these are not more harassing to the subject, than they are disparaging to the truth and authority of the superior. It is in the bosom of a well-conducted family, where you witness the sure dispensation of all the reward and encouragement which have been promised, and the unfailing execution of the disgrace and the dismissal that are held forth to obstinate disobedience. Now those very qualities of which this uniformity is the test and the characteristic in the government of any human society, of these also is it the test and the characteristic in the government of Nature. It bespeaks the wisdom, and the authority, and the truth of Him who framed and who administers. Let there be a King eternal, immortal, and invisible, and let this universe be His empire—and in all the rounds of its complex but unerring mechanism, do I recognize him as the only wise God. In the constancy of Nature, do I read the constancy and truth of that great master Spirit, who hath imprinted His own character on all that hath emanated from His power; and when told that throughout the mighty lapse of centuries, all the courses both of earth and of heaven, have been upholden as before, I only recognize the footsteps of Him who is ever the same, and whose faithfulness is unto all generations. That perpetuity, and order, and ancient law of succession, which have subsisted so long, throughout the wide diversity of things, bear witness to the Lord of hosts, as still at the head of His well marshalled family. The present age is only re-echoing the lesson of all past ages—and that spectacle, which has misled those who by wisdom know not God, into dreary atheism, has enhanced every demonstration both of his veracity and power, to all intelligent worshippers. We know that all things continue as they were from the beginning of creation. We know that the whole of surrounding materialism stands forth, to this very hour, in all the inflexibility of her wonted characters. We know that heaven, and earth, and sea, still discharge the same functions, and subserve the very same beneficent processes. We know that astronomy plies the same rounds as before, that the cycles of the firmament move in their old and appointed order, and that the year circulates, as it has ever done, in grateful variety, over the face of an expectant world—but only because all are of God, and they continue this day according to His ordinances—for all are His servants.

Now it is just because the successions which take place in the economy of Nature, are so invariable, that we should expect the successions which take place in the economy of God's moral government to be equally invariable. That expectation which He never disappoints when it is the fruit of a universal instinct, He surely will never disappoint when it is the fruit of His own express and immediate revelation. If because God hath so established it, it cometh to pass, then of whatsoever it may be affirmed that God hath so said it, it will come equally to pass. I should certainly look for the same character in the administrations of His special grace, that I, at all times, witness in the administrations of His ordinary providence. If I see in the system of His world, that the law by which two events follow each other, gives rise to a connection between them that never is dissolved, then should He say in his word, that there are certain invariable methods of succession, in virtue of which, when the first term of it occurs, the second is sure at all times to follow, I should be very sure in my anticipations, that it will indeed be most punctually and most rigidly so. It is thus, that the constancy of Nature is in fullest harmony with the authority of Revelation—and that, when fresh from the contemplation of the one, I would listen with most implicit faith to all the announcements of the other.

When we behold all to be so sure and settled in the works of God, then may we look for all being equally sure and settled in the word of God. Philosophy hath never yet detected one iota of deviation from the ordinances of Nature—and never, therefore, may we conclude, shall the experience either of past or future ages, detect one iota of deviation from the ordinances of Revelation. He who so pointedly adheres to every plan that He hath established in creation, will as pointedly adhere to every proclamation that He hath uttered in Scripture. There is nought of the fast and loose in any of His processes—and whether in the terrible denunciations of Sinai, or those mild proffers of mercy that were sounded forth upon the world through Messiah, who upholdeth all things by the word of His power, shall we alike experience that God is not to be mocked, and that with Him there is no variableness, neither shadow of turning.

With this certainty, then, upon our spirits, let us now look not to the successions which He hath instituted in Nature, but to the successions which he hath announced to us in the word of His testimony—and let us, while so doing, fix and solemnize our thoughts by the consideration, that as God hath said it, so will He do it.

The first of these successions, then, on which we may count infallibly, is that which He hath proclaimed between sin and punishment. The soul that sinneth it shall die. And here there is a common ground on which the certainties of divine revelation meet and are at one with the certainties of human experience. We are told in the Bible, that all have sinned, and that, therefore, death hath passed upon all men. The connection between these two terms is announced in Scripture to be invariable—and all observation tells us, that it is even so. Such was the sentence uttered in the hearing of our first parents; and all history can attest how God hath kept by the word of His threatening—and how this law of jurisprudence from heaven is realized before us upon earth, with all the certainty of a law of Nature. The death of man is just as stable and as essential a part of his physiology, as are his birth, or his expansion, or his maturity, or his decay. It looks as much a thing of organic necessity, as a thing of arbitrary institution—and here do we see blended into one exhibition, a certainty of the divine word that never fails, and a constancy in Nature that never is departed from. It is indeed a striking accordancy that what in one view of it appears to be a uniform process of Nature, in another view of it, is but the unrelenting execution of a dread utterance from the God of Nature. From this contemplation may we gather, that God is as certain in all His words, as he is constant in all His ways. Men can philosophize on the diseases of the human system—and the laborious treatise can be written on the class, and the character, and the symptoms, of each of them—and in our halls of learning, the ample demonstration can be given, and disciples may be taught how to judge and to prognosticate, and in what appearances to read the fell precursors of mortality—and death has so taken up its settled place among the immutabilities of Nature, that it is as familiarly treated in the lecture-rooms of science, as any other phenomena which Nature has to offer for the exercise of the human understanding. And, O how often are the smile and the stoutness of infidelity seen to mingle with this appalling contemplation—and how little will its hardy professors bear to be told, that what gives so dread a certainty to their speculation is, that the God of Nature and the God of the Bible, are one—that when they describe, in lofty nomenclature, the path of dying humanity, they only describe the way in which He fulfils upon it His irrevocable denunciation—that He is but doing now to the posterity of Adam what He told to Adam himself on his expulsion from paradise—and that, if the universality of death prove how every law in the physics of creation is sure, it just as impressively proves, how every word of God's immediate utterance to man, or how every word of prophecy is equally sure.

And in every instance of mortality which you are called to witness, do we call upon you to read in it the intolerance of God for sin, and how unsparingly and unrelentingly it is, that God carries into effect His every utterance against it. The connection which He hath instituted between the two terms of sin and of death, should lead you from every appeal that is made to your senses by the one, to feel the force of an appeal to your conscience by the other. It proves the hatefulness of sin to God, and it also proves with what unfaltering constancy God will prosecute every threat, until He hath made an utter extirpation of sin from His presence. There is nought which can make more palpable the way in which God keeps every saying in His perpetual remembrance, and as surely proceeds upon it, than doth this universal plague wherewith He hath smitten every individual of our species, and carries off its successive generations from a world that sprung from His hand in all the bloom and vigour of immortality. When death makes entrance upon a family, and, perhaps, seizes on that one member of it, all whose actual transgressions might be summed up in the out breakings of an occasional waywardness, wherewith the smiles of infant gaiety were chequered—still how it demonstrates the unbending purposes of God against our present accursed nature, that in some one or other of its varieties, every specimen must die. And so it is, that from one age to another, He makes open manifestation to the world, that every utterance which hath fallen from him is sure; and that ocular proof is given to the character of Him who is a Spirit, and is invisible; and that sense lends its testimony to the truth of God, and the truth of His Scripture; and that Nature, when rightly viewed, instead of placing its inquirers at atheistical variance with the Being who upholds it, holds out to us the most impressive commentary that can be given, on the reverence which is due to all His communications, even by demonstrating, that faith in His word is at unison with the findings of our daily observation.

But God hath further said of sin and of its consequences, what no observation of ours has yet realized. He hath told us of the judgment that cometh after death, and He hath told us of the two diverse paths which lead from the judgment-seat unto eternity. Of these we have not yet seen the verification, yet surely we have now seen enough to prepare us for the unfailing accomplishment of every utterance that cometh from the lips of God. The unexcepted death which we know cometh upon all men, for that all have sinned, might well convince us of the certainty of that second death which is threatened upon all who turn not from sin unto the Saviour. There is an indissoluble succession here between our sinning and our dying—and we ought now to be so aware of God as a God of precise and peremptory execution, as to look upon the succession being equally indissoluble, between our dying in sin now, and rising to everlasting condemnation hereafter. The sinner who wraps himself in delusive security—and that, because all things continue as they have done, does not reflect of this very characteristic, that it is indeed the most awful proof of God's immutable counsels, and to himself the most tremendous presage of all the ruin and wretchedness which have been denounced upon him. The spectacle of uniformity that is before his eyes, only goes to ascertain that as God hath purposed, so, without vacillation or inconstancy, will he ever perform. He hath already given a sample, or an earnest of this, in the awful ravages of death; and we ask the sinner to behold, in the ever-recurring spectacle of moving funerals, and desolated families, the token of that still deeper perdition which awaits him. Let him not think that the God who deals His relentless inflictions here on every son and daughter of the species, will falter there from the work of vengeance that shall then descend on the heads of the impenitent. O how deceived then are all those ungodly, who have been building to themselves a safety and an exemption on the perpetuity of Nature! All the perpetuity which they have witnessed, is the pledge of a God who is unchangeable—and who, true to His threatening as to every other utterance which passes his lips, hath said, in the hearing of men and of angels, that the soul which is in sin shall perish.

But, secondly, there is another succession announced to us in Scripture, and on the certainty of which we may place as firm a reliance as on any of the observed successions of Nature—even that which obtains between faith and salvation. He who believeth in Christ, shall not perish, but shall have life everlasting. The same truth which God hath embarked on the declarations of His wrath against the impenitent, He hath also embarked on the declarations of His mercy to the believer. There is a law of continuity, as unfailing as any series of events in Nature, that binds with the present state of an obstinate sinner upon earth, all the horrors of his future wretchedness in hell—but there is also another law of continuity just as unfailing, that binds the present state of him who putteth faith in Christ here, with the triumphs and the transports of his coming glory hereafter. And thus it is, that what we read of God's constancy in the book of Nature, may well strengthen our every assurance in the promises of the Gospel. It is not in the recurrence of winter alone, and its desolations, that God manifests His adherence to established processes. There are many periodic evolutions of the bright and the beautiful along the march of His administrations—as the dawn of morn; and the grateful access of spring, with its many hues, and odours, and melodies; and the ripened abundance of harvest; and that glorious arch of heaven, which Science hath now appropriated as her own, but which nevertheless is placed there by God as the unfailing token of a sunshine already begun, and a storm now ended—all these come forth at appointed seasons, in a consecutive order, yet mark the footsteps of a beneficent Deity. And so the economy of grace has its regular successions, which carry however a blessing in their train. The faith in Christ, to which we are invited upon earth, has its sure result, and its landing-place in heaven—and just with as unerring certainty as we behold in the courses of the firmament, will it be followed up by a life of virtue, and a death of hope, and a resurrection of joyfulness, and a voice of welcome at the judgment-seat, and a bright ascent into fields of ethereal blessedness, and an entrance upon glory, and a perpetual occupation in the city of the living God.

To all men hath He given a faith in the constancy of Nature, and He never disappoints it. To some men hath He given a faith in the promises of the Gospel, and He is ready to bestow it upon all who ask, or to perfect that which is lacking in it—and the one faith will as surely meet with its corresponding fulfilment as the other. The invariableness that reigns throughout the kingdom of Nature, guarantees the like invariableness in the kingdom of grace. He who is steadfast to all His appointments, will be true to all His declarations—and those very exhibitions of a strict and undeviating order in our universe, which have ministered to the irreligion of a spurious philosophy, form a basis on which the believer can prop a firmer confidence than before, in all the spoken and all the written testimonies of God.

With a man of taste, and imagination, and science, and who is withal a disciple of the Lord Jesus, such an argument as this must shed a new interest and glory over His whole contemplation of visible things. He knows of His Saviour, that by Him all things were made, and that by Him too all things are upholden. The world, in fact, was created by that Being whose name is the Word; and from the features that are imprinted on the one, may he gather some of the leading characteristics of the other. More expressly will he infer from that sure and established order of Nature, in which the whole family of mankind are comprehended, that the more special family of believers are indeed encircled within the bond of a sure and a well-ordered covenant. In those beauteous regularities by which the one economy is marked, will he be led to recognise the "yea" and the "amen" which are stamped on the other economy—and when he learns that the certainties of science are unfailing, does he also learn that the sayings of Scripture are un alterable. Both he knows to emanate from the same source; and every new experience of Nature's constancy, will just rivet him more tenaciously than before to the doctrine and the declarations of his Bible. Furnished with such a method of interpretation as this, let him go abroad upon Nature, and all that he sees will heighten and establish the hopes which Revelation hath awakened. Every recurrence of the same phenomena as before, will be to him a distinct testimony to the faithfulness of God. The very hours will bear witness to it. The lengthening shades of even will repeat the lesson held out to him by the light of early day—and when night unveils to his eye the many splendours of the firmament, will every traveller on his circuit there, speak to him of that mighty and invisible King, all whose ordinations are sure. And this manifestation from the face of heaven, will be reflected to him by the panorama upon earth. Even the buds which come forth at their appointed season on the leafless branches; and the springing up of the flowers and the herbage, on the spots of ground from which they had disappeared; and that month of vocal harmony wherewith the mute atmosphere is gladdened as before, with the notes of joyous festival; and so, the regular march of the advancing year through all its footsteps of revival, and progress, and maturity, and decay—these are to him but the diversified tokens of a God whom he can trust, because of a God who changeth not. To his eyes, the world reflects upon the word the lesson of its own wondrous harmony; and his science, instead of a meteor that lures from the greater light of Revelation, serves him as a pedestal on which the stability of Scripture is more firmly upholden.

The man who is accustomed to view aright the uniformity of Nature's sequences, will be more impressed with the certainty of that sequence, which is announced in the Bible between faith and salvation—and he of all others, should re-assure his hopes of immortality, when he reads, that the end of our faith is the salvation of our souls. In this secure and wealthy place, let him take up his rest, and rejoice himself greatly with that God who has so multiplied upon him the evidences of His faithfulness. Let him henceforth feel that he is in the hands of one who never deviates, and who cannot lie—and who, as He never by one act of caprice, hath mocked the dependence that is built on the foundation of human experience, so, never by one act of treachery, will He mock the dependence that is built on the foundation of the divine testimony. And more particularly, let him think of Christ, who hath all the promises in His hand, that to Him also all power has been committed in heaven and in earth —and that presiding therefore, as he does, over that visible administration, of which constancy is the unfailing attribute, He by this hath given us the best pledge of a truth that abideth the same, to day, and yesterday, and for ever.

We are aware, that no argument can of itself work in you the faith of the Gospel—that words and reasons, and illustrations, may be multiplied without end, and yet be of no efficacy—that if the simple manifestation of the Spirit be withheld, the expounder of Scripture, and of all its analogies with Creation or Providence, will lose his labour —and while it is his part to prosecute these to the uttermost, yet nought will he find more surely and experimentally true, than that without a special interposition of light from on high, he runneth in vain, and wearieth himself in vain. It is for him to ply the instrument, it is for God to give unto it the power which availeth. We are told of Christ on His throne of mediatorship, that He hath all the energies of Nature at command, and up to this hour do we know with what a steady and unfaltering hand He hath wielded them. Look to the promise as equally steadfast, of "Lo, I am with you always, even unto the end of the world"—and come even now to His own appointed ordinance in the like confidence of a fellowship with Him, as you would to any of the scenes or ordinations of Nature, and in the confidence that there the Lord of Nature will prove Himself the same that He has ever been.* The blood that was announced many centuries ago to cleanse from all sin, cleanseth still. The body which hath borne in all past ages the iniquity of believers, beareth it still. That faith which appropriates Christ and all the benefits of His purchase, to the soul, still performs the same office. And that magnificent economy of Nature which was established at the first, and so abideth, is but the symbol of that higher economy of grace which continueth to this day according to all its ordinances.

* This Sermon was delivered on the morning of a Communion Sabbath.

"Whosoever eateth my flesh, and drinketh my blood," says the Saviour, "shall never die." When you sit down at His table, you eat the bread, and you drink the wine by which these are represented —and if this be done worthily, if there be a right correspondence between the hand and the heart in this sacramental service, then by faith do you receive the benefits of the shed blood, and the broken body; and your so doing will as surely as any succession takes place in the instituted courses of Nature, be followed up by your blessed immortality. And the brighter your hopes of glory hereafter, the holier will you be in all your acts and affections here. The character even now will receive a tinge from the prospect that is before you—and the habitual anticipation of heaven will bring down both of its charity and its sacredness upon your heart. He who hath this hope in him purifieth himself even as Christ is pure—and even from the present if a true approach to the gate of His sanctuary, will you carry a portion of His spirit away with you. In partaking of these His consecrated elements, you become partakers of his gentleness and devotion, and unwearied beneficence—and because like Him in time, you will live with Him through eternity.

DISCOURSE II.

ON THE CONSISTENCY BETWEEN THE EFFICACY OF PRAYER—AND THE UNIFORMITY OF NATURE.

"Knowing this first, that there shall come in the last days scoffers, walking after their own lusts,—and saying, Where is the promise of his coming? for since the fathers fell asleep, all things continue as they were from the beginning of the creation."—2 PETER iii. 3, 4.

THE infidelity spoken of in our text, had for its basis the stability of nature, or rested on the imagination that her economy was perpetual and everlasting—and every day of nature's continuance added to the strength and inveteracy of this delusion. In proportion to the length of her past endurance, was there a firm confidence felt in her future perpetuity. The longer that nature lasted, or the older she grew, her final dissolution was held to be all the more improbable—till nothing seemed so unlikely to the atheistical men of that period, as the intervention of a God with a system of visible things, which looked so unchanging and so indestructible. It was like the contest of experience and faith, in which the former grew every day stronger and stronger, and the latter weaker and weaker, till at length it was wholly extinguished; and men in the spirit of defiance or ridicule, braved the announcement of a Judge who should appear at the end of the world, and mocked at the promise of His coming.

But there is another direction which infidelity often takes, beside the one specified in our text. It not only perverts to its own argument, what experience tells of the stability of nature; and so concludes that we have nothing to fear from the mandate of a God, laying sudden arrest and termination on its processes. It also perverts what experience tells of the uniformity of nature; and so concludes that we have nothing either to hope or to fear, from the intervention of a God during the continuance or the currency of these processes. Beside making nature independent of God for its duration, which they hold to be everlasting; they would also make nature to be independent of God for its course, which they hold to be unalterable. They tell us of the rigid and undeviating constancy from which nature is never known to fluctuate; and that in her immutable laws, in the march and regularity of her orderly progressions, they can discover no trace whatever of any interposition by the finger of a Deity. It is not only that all things continue to be, as they were from the beginning of creation; but that all things continue to act, as they did from the beginning of the creation—causes and effects following each other in wonted and invariable succession, and the same circumstances ever issuing in the same consequents as before. With such a system of things, there is no room in their creed or in their imagination, for the actings of a God. To their eye, nature proceeds by the sure footsteps of a mute and unconscious materialism; nor can they recognize in its evolutions those characters of the spontaneous or the wilful, which bespeak a living God to have had any concern with it. He may have formed the mundane system at the first: he may have devised for matter its properties and its laws: but these properties, they tell us, never change; these laws never are relaxed or receded from. And so we may as well bid the storm itself cease from its violence, as supplicate the unseen Being whom we fancy to be sitting aloft and to direct the storm. This they hold to be a superstitious imagination, which all their experience of nature and of nature's immutability forbids them to entertain. By the one infidelity, they have banished a God from the throne of judgment. By the other infidelity, they have banished a God from the throne of providence. By the first they tell us, that a God has nought to do with the consummation of nature; or, rather, that nature has no consummation. By the second, they tell us that a God has nought to do with the history of nature. The first infidelity would expunge from our creed the doctrine of a coming judgment. The second would expunge from it the doctrine of a present and a special providence, and the doctrine of the efficacy of prayer.

Now this last, though not just the infidelity of the text—yet being very much the same with it in principle—we hold it sufficiently textual, though we make it, and not the other the subject of our present argument. We admit the uniformity of visible nature—a lesson forced upon us by all experience. We admit that as far as our observation extends, nature has always proceeded in one invariable order—insomuch that the same antecedents have, without exception, been ever followed up by the same consequents; and that, saving the well accredited miracles of the Jewish and Christian dispensations, all things have so continued since the beginning of the creation.

We admit that, never in our whole lives, have we witnessed as the effect of man's prayer, any infringement made on the known laws of the universe; or that nature by receding from her constancy, to the extent that we have discovered it, has ever in one instance yielded to his supplicating cry. We admit that by no importunity from the voice of faith, or from any number and combination of voices, have we seen an arrest or a shift laid on the ascertained courses, whether of the material or the mental economy; or a single fulfilment of any sort, brought about in contravention, either to the known properties of any substance, or to the known principles of any established succession in the history of nature. These are our experiences; and we are aware the very experiences which ministered to the infidelity of our text, and do minister to the practical infidelity of thousands in the present day—yet, in opposition to, or rather notwithstanding these experiences, universal and unexcepted though they be, do we affirm the doctrine of a superintending providence, as various and as special, as our necessities—the doctrine of a perpetual interposition from above, as manifoldly and minutely special, as are the believing requests which ascend from us to Heaven's throne.

We feel the importance of the subject, both in

its application to the judgment that now hangs over us,* and to the infidelity of the present times. But we cannot hope to be fully understood, without your most strenuous and sustained attention—an attention, however, which we request may be kept up to the end, even though certain parts in the train of observation may not have been followed by you. What some may lose in those passages, where the subject is presented in the form of a general argument, may again be recovered, when we attempt to establish our doctrine by scripture, or to illustrate it by instances taken from the history of human affairs. In one way or other, you may seize on the reigning principle of that explanation, by which we endeavour to reconcile the efficacy of prayer with the uniformity of experience. And our purpose shall have been obtained, if we can at all help you to a greater confidence in the reality of a superintending providence, to a greater comfort and confidence in the act of making your requests known unto God.

Let us first give our view in all its generality, in the hope that any obscurity which may rest upon it in this form, will be dissipated or cleared up, in the subsequent appeals that we shall make, both to the lessons of the Bible, and to the lessons of human experience.

We grant then, we unreservedly grant, the uniformity of visible nature; and now let us compute how much, or how little, it amounts to. Grant of all our progressions, that, as far as our eye can carry us, they are invariable; and then let us only

* This sermon was preached during the prevalence of cholera.

reflect how short a way we can trace any of them upwards. In speculating on the origin of an event, we may be able to assign the one which immediately preceded, and term it the proximate cause; or even ascend by two or three footsteps, till we have discovered some anterior event which we term the remote cause. But how soon do we arrive at the limit of possible investigation, beyond which if we attempt to go, we lose ourselves among the depths and the obscurities of a region that is unknown? Observation may conduct us a certain length backwards in the train of causes and effects; but, after having done its uttermost, we feel, that, above and beyond its loftiest place of ascent, there are still higher steps in the train, which we vainly try to reach, and find them inaccessible. It is even so throughout all philosophy. After having arrived at the remotest cause which man can reach his way to, we shall ever find there are higher and remoter causes still, which distance all his powers of research, and so will ever remain in deepest concealment from his view. Of this higher part of the train he has no observation. Of these remoter causes, and their mode of succession, he can positively say nothing. For aught he knows, they may be under the immediate control of higher beings in the universe; or, like the upper part of a chain, a few of whose closing links are all that is visible to us, they may be directly appended to the throne, and at all times subject to the instant pleasure of a prayer-hearing God. And it may be by a responsive touch at the higher, and not the lower part of the progression, that He answers our prayers. It may be not by an act of intervention among those near and visible causes, where intervention would be a miracle; it may be by an unseen, but not less effectual act of intervention, among the remote and therefore the occult causes, that He adapts Himself to the various wants and meets the various petitions of His children. If it be in the latter way that He conducts the affairs of His daily government—then may He rule by a providence as special, as are the needs and the occasions of His family; and, with an ear open to every cry, might He provide for all and administer to all, without one infringement on the uniformity of visible nature. If the responsive touch be given at the lower part of the chain, then the answer to prayer is by miracle, or by a contravention to some of the known sequences of nature. But if the responsive touch be given at a sufficiently higher part of the chain, then the answer is as effectually made, but not by miracle, and without violence to any one succession of history or nature which philosophy has ascertained—because the reaction to the prayer strikes at a place that is higher than the highest investigations of philosophy. It is not by a visible movement within the region of human observation, but by an invisible movement in the transcendental region above it, that the prayer is met and responded to. The Supernal Power of the Universe, the mighty and unseen Being who sits aloft, and has been significantly styled the Cause of causes—He, in immediate contact with the upper extremities of every progression, there puts forth an overruling influence which tells and propagates downwards to the lower extremities; and so, by an agency placed too remote either for the eye of sense or for all the instruments of science to discover, may God, in answer if He choose to prayer, fix and determine every series of events—of which nevertheless all that man can see is but the uniformity of the closing footsteps—a few of the last causes and effects following each other in their wonted order. It is thus that we reconcile all the experience which man has of nature's uniformity, with the effect and significancy of his prayers to the God of nature. It is thus that at one and the same time, do we live under the care of a presiding God, and among the regularities of a harmonious universe.

These views are in beautiful accordance with the simple and sublime theology unfolded to us in the book of Job—where, whether in the movements of the animated kingdom below, or the great evolutions that take place in the upper regions of the atmosphere, the phenomena and the processes of visible nature are sketched with a masterly hand. It is in the midst of these scenes and impressive descriptions, that we are told—"lo these are parts of his ways." The translation does not say what parts; but the original does. They are but the lower parts—the endings as it were of the different processes—the last and lowest footsteps, which are all that science can investigate; and of which, throughout the whole of her limited ascent, she has traced the uniformity. But she has traced it a very short way: or, in the language of the patriarch, who estimates aright the achievements of philosophy—how little a portion is heard of Him—how few the known footsteps which are beneath the veil to the unknown steps and workings which are above it; and so, the thunder, or rather the inward and secret movements of His power, who can understand?

"He bindeth up the waters in his thick clouds: and the cloud is not rent under them. He holdeth back the face of his throne, and spreadeth his cloud upon it. He hath compassed the waters with bounds, until the day and night come to an end. The pillars of heaven tremble, and are astonished at his reproof. He divideth the sea with his power, and by his understanding he smiteth through the proud. By his spirit he hath garnished the heavens; his hand hath formed the crooked serpent. Lo, these are parts of his ways; but how little a portion is heard of him? but the thunder of his power who can understand?" Job xxvi. 8—14.

The last sentence of this magnificent passage were better translated thus—These are the parts, or the lower endings of his ways;—but the secret working of his power, who can understand?

That part of the economy of the divine administration, in virtue of which God works, not without but by secondary causes, is frequently intimated in the book of Psalms.

"Who maketh his angels spirits, his ministers a flaming fire." Ps. civ. 4.

Or, as it might have been translated—"Who maketh the winds his messengers, and the flaming fire his servant."

But without the aid of any emendations in our version, this subserviency of visible nature to the invisible God, is distinctly laid before us in the following passages.

"They that go down to the sea in ships, that do business in great waters; These see the works of the Lord, and his wonders in the deep. For he commandeth, and raiseth the stormy wind, which lifteth up the waves thereof. They mount up to the heaven, they go down again to the depths; their soul is melted because of trouble. They reel to and fro, and stagger like a drunken man, and are at their wit's end. Then they cry unto the Lord in their trouble, and he bringeth them out of their distresses. He maketh the storm a calm, so that the waves thereof are still. Then are they glad, because they be quiet; so he bringeth them unto their desired haven. Oh, that men would praise the Lord for his goodness, and for his wonderful works to the children of men." Psalm cvii. 23—31.

He raises the tempest, not without the wind, but by the wind. In the one way, it would have been a miracle; in the other way it is alike effectual, but without any change in the properties or laws of visible nature—without what we commonly understand by a miracle. He does not bring the vessel against the wind to its desired haven; but he makes the storm a calm, and so the waves thereof are still. Our Saviour also bade the winds into peace; and the miracle there lay in the effect following on the heard utterance of His voice. A voice no less effectual though unheard by us, overrules at all times the working of nature's elements; and brings the ordinary processes, as well as the marked and miraculous exception to them, under the control of a divine agency.

"Whatsoever the Lord pleased, that did he in heaven, and in earth, in the seas, and all deep places. He causeth the vapours to ascend from the ends of the earth; he maketh lightnings for the rain; he bringeth the wind out of his treasuries." Psalm cxxxv. 6, 7.

Here, without any change of translation, we are told of the subserviency of the visible instruments, to the invisible but real agency of Him who wields them at His pleasure. In this passage, the winds are plainly represented to us as the messengers of God, and the flaming fire as his servant. He changes no properties, and no visible processes—working, not without the wind, but by it—not without the electric matter, but by it—not without the rain, but by it—not without the vapour, but by it. Let the philosopher tell how far back he can go, in exploring the method and order of these respective agencies. Then we have only to point further back and ask—on what evidence he can tell, that the fiat and the finger of a God are not there. We grant the observed order to be invariable, save when God chooses to interpose by miracle. But whether he does or not—from that chamber of his hidden operations, which philosophy has not found its way to, can he so direct all, so subordinate all, that whatever the Lord pleases, that does he in heaven and in earth, in the seas, and all deep places.

"Praise the Lord from the earth, ye dragons, and all deeps: Fire and hail; snow and vapour; stormy wind fulfilling his word." Psalm cxlviii. 7, 8.

The stormy wind fulfilleth his word.

Our last example shall be from the New Testament. "Nevertheless he left not himself without witness, in that he did good, and gave us rain from heaven, and fruitful seasons, filling our hearts with food and gladness." Acts xiv. 17.

This last example will prepare you to go along with one of the particular instances we are just to bring forward, of a special prayer met by a special fulfilment.

We are thus enabled to perceive what the respective provinces are of philosophy and faith. Every event in nature or history, has a cause in some prior event that went before it, and that again in another, and that again in another still higher than itself in this scale of precedency; and so might we climb our ascending way from cause to cause, from consequent to antecedent—till the investigation has been carried upwards, from the farthest possible verge of human discovery. There it is that the domain of observation or of philosophy terminates; but we mistake, if we think that there the progression, whose terms or whose footsteps we have traced thus far, also terminates. Beyond this limit we cannot track the pathway of causation—not because the pathway ceases, but because we have lost sight of it—having now retired from view among the depths and mysteries of an unknown region, which we, with our bounded faculties, cannot enter. This may be termed the region of faith—placed as it were above the region of experience. The things which are done in the higher, have an overruling influence, by lines of transmission, on all that happens in the lower—yet without one breach or interruption to the uniformity of visible nature. Whatever is done in the transcendental region—be it by the influence of prayer; by the immediate finger of God; by the ministry of angels; by the spontaneous movements, whether of displeasure or of mercy above, responding to the sins or to the supplicating cries that ascend from earth's inhabitants below—that will pass by a descending influence into the palpable region of sense and observation—yet, from the moment it comes within its limits, will it proceed without the semblance of a miracle, but by the march and the movement of nature's regularity, to its final consummation. God hath in wisdom ordained a regimen of general laws; and, that man might gather from the memory of the past those lessons of observation which serve for the guidance of the future, He hath enacted that all those successions shall be invariable, which have their place and their fulfilment within the world of sensible experience. Yet God has not, on that account, made the world independent of Himself. He keeps a perpetual hold on all its events and processes notwithstanding. He does not dissever Himself, for a single instant, from the government and the guardianship of His own universe; and can still, notwithstanding all we see of nature's rigid uniformity, adapt the forth-goings of His power to all the wants and all the prayers of His dependent family. For this purpose, He does not need to stretch forth His hand on the inferior and the visible links of any progression, so as to shift the known successions of experience; or at all to intermeddle with the lessons and the laws of this great schoolmaster. He may work in secret, and yet perform all His pleasure—not by the achievement of a miracle on nature's open platform; but by the touch of one or other of those master springs, which lie within the recesses of her inner laboratory. There, and at His place of supernal command by the fountain heads of influence, He can turn whithersoever He will the machinery of our world, and without the possibility by human eye of detecting the least infringement on any of its processes—at once upholding the regularity of visible nature, and the supremacy of nature's invisible God.

But we are glad to make our escape, and now to make it conclusively, from the obscurer part of our reasoning on this subject—although, most assuredly, these are not the times for passing it wholly by; or for withholding aught which can make in favour of the much derided cause of humble and earnest piety. But, instead of propounding our doctrine in the terms of a general argument, let us try the effect of a few special instances—by which, perhaps, we might more readily gain the consent of your understanding to our views.

When the sigh of the midnight storm sends fearful agitation into a mother's heart, as she thinks of her sailor boy, now exposed to its fury, on the waters of a distant ocean—these stern disciples of a hard and stern infidelity would, on this notion of a rigid and impracticable constancy in nature, forbid her prayers—holding them to be as impotent and vain, though addressed to the God who has all the elements in his hand, as if lifted up with senseless importunity to the raving elements themselves. Yet nature would strongly prompt the aspiration; and, if there be truth in our argument, there is nothing in the constitution of the universe to forbid its accomplishment. God might answer the prayer, not by unsettling the order of secondary causes—not by reversing any of the wonted successions that are known to take place, in the ever-restless ever-heaving atmosphere—not by sensible miracle among those nearer footsteps which the philosopher has traced; but by the touch of an immediate hand among the deep recesses of materialism, which are beyond the ken of all his instruments. It is thence that the Sovereign of nature might bid the wild uproar of the elements into silence. It is there that the virtue comes out of Him, which passes like a winged messenger from the invisible to the visible; and, at the threshold of separation between these two regions, impresses the direction of the Almighty's will on the remotest cause which science can mount her way to. From this point in the series, the path of descent along the line of nearer and proximate causes may be rigidly invariable; and in respect of the order, the precise undeviating order, wherewith they follow each other, all things continue as they were from the beginning of the creation. The heat, and the vapour, and the atmospherical precipitates, and the consequent moving forces by which either to raise a new tempest or to lay an old one—all these may proceed, and without one hair-breadth of deviation, according to the successions of our established philosophy—yet each be but the obedient messenger of that voice, which gave forth its command at the fountain-head of the whole operation; which commissioned the vapours to ascend from the ends of the earth, and made lightnings for the rain, and brought the wind out of his treasuries. These are the palpable steps of the process; but an unseen influence, behind the farthest limit of man's boasted discoveries, may have set them agoing. And that influence may have been accorded to prayer—the power that moves Him, who moves the universe; and who, without violence to the known regularities of nature, can either send forth the hurricane over the face of the deep or recall it at His pleasure. Such is the joyful persuasion of faith, and proud philosophy cannot disprove it. A woman's feeble cry may have overruled the elemental war; and hushed into silence this wild frenzy of the winds and the waves; and evoked the gentler breezes from the cave of their slumbers; and wafted the vessel of her dearest hopes, and which held the first and fondest of her earthly treasures, to its desired haven.

And so of other prayers. It is not without instrumentality, but by means of it, that they are answered. The fulfilment is preceded by the accustomed series of causes and effects; and preceded as far upward, as the eye of man can trace the pedigree of sensible causation. Were it by a break anywhere in the traceable part of this series that the prayer was answered, then its fulfilment would be miraculous. But without a miracle the prayer is answered as effectually. Thus, for example, is met the cry of a people under famine, for a speedy and plenteous harvest—not by the instant appearance of the ripened grain, at the bidding of a voice from heaven—not preternaturally cherished into maturity, in the midst of storms; but ushered onwards, by a grateful succession of shower and sunshine, to a prosperous consummation. An abundant harvest is granted to prayer—yet without violence, either to the laws of the vegetable physiology, or to any of the known laws by which the alterations of the weather are determined. It must be acknowledged by every philosopher, how soon it is that we arrive in both departments on the confines of deepest mystery: and, let the constancy of patent and palpable nature be as unaltered and unalterable as it may, God reserves to Himself the place of mastery and command, whether among the arcana of vegetation or the depths of meteorology, He may at once permit a most rigid uniformity to the visible workings of nature's mechanism—while among its invisible, which are also its antecedent workings, He retains that station of preeminence and power, whence He brings all things to pass according to His pleasure. It is not by sending bread from the upper storehouses of the firmament, that He answers this prayer. It is by sending rain and fruitful seasons. The intermediate machinery of nature is not cast aside, but pressed into the service; and the prayer is answered by a secret touch from the finger of the Almighty, which sets all its parts and all its processes agoing. With the eye of sense, man sees nothing but nature revolving in her wonted cycles, and the months following each other in bright and beautiful succession. In the eye of faith, ay and of sound philosophy, every year of smiling plenty upon earth is a year crowned with the goodness of heaven.

But to touch on that which more immediately concerns us, let us now instance prayer for health. We ask, if here philosophy has taken possession of the whole domain, and left no room for the prerogatives and the exercise of faith—no hope for prayer? Has the whole intermediate space between the first cause and the ultimate phenomena, been so thoroughly explored; and the rigid uniformity of every footstep in the series been so fixed and ascertained by observation, as to preclude the rationality of prayer, and leave it without a meaning, because without the possibility of a fulfilment? Where is the physician or the physiologist who can tell, that he has made the ascent from one prognostic or one predisposition to another—till he reached even to the primary fountain-head of that influence, which either medicates or distempers the human frame; and found throughout an adamantine chain of necessity, not to be broken by the sufferer's imploring cry? We ask the guardians of our health, how far upon the pathway of causation, the discoveries of medical science have carried them; and whether, above and beyond their farthest look into the mysteries of our framework, there are not higher mysteries; where a God may work in secret, and the hand of the Omnipotent be stretched forth to heal or to destroy? It is thence, He may answer prayer. It is from this summit of ascendancy, that He may direct all the processes of the human constitution—yet without violating in any instance, the uniformity of the few last and visible footsteps. Because science has traced, and so far determined this uniformity, she has not therefore exiled God from His own universe: She has not forced the Deity to quit His hold of its machinery, or to forego by one iota the most perfect command of all its evolutions. His superintendence is as close and continuous and special, as if all things were done by the visible intervention of his hand. Without superstition, with the fullest recognition of science in all its prerogatives and all its glories—might we feel our immediate dependence on God; and, even in this our philosophic day, and notwithstanding all that philosophy has made known to us, might we still assert and vindicate the higher philosophy of prayer—asking of God, as patriarchs and holy men of old did before us, for safety and sustenance and health and all things.

And if ever in the dealings of God with the people of the earth, if ever science had less of the territory and faith had more of it, it is in that undisclosed mystery which still hangs over us; which now for many months has shed baleful influences on your crowded city; and whereof no man can tell whether in another day or another hour, it might not descend with fell swoop into the midst of his own family—entering there with rude unceremonious footstep, and hurrying to one of its rapid and inglorious funerals the dearest of the inmates. Never on any other theme did philosophy make more entire demonstration of her own helplessness; and perhaps at the very first footstep of the investigation, or on the question of the proximate cause, the controversy is loudest of all. But however justly of the proximate cause discovery may be made, or however remotely among the anterior causes the investigation might be carried, never will proud philosophy be able to annul the intervention of a God, or purchase to herself the privilege of mocking at the poor man's prayer. Indeed, amid the exuberance and variety of speculation on this unsettled and unknown subject, there was one remote cause assigned for this pestilent visitation, which, so far from shutting out, rather suggests and that most forcibly the intervention of a God immediately before it. "And it shall come to pass in that day, that the Lord shall hiss for the fly that is in the uttermost part of the rivers of Egypt, and for the bee that is in the land of Assyria: and they shall come, and shall rest all of them in the desolate valleys, and in the holes of the rocks, and upon all thorns, and upon all bushes."* We hope to have made it plain to you, let this or any other cause be found the true one, that, however high the path of discovery may have been traced, yet higher still there is place for the finger of a God above to regulate all

* Isaiah vii. 18, 19.

the designs of a special providence, and to move in conformity with all the accepted prayers of His family below. But among the scoffers of our latter day, even in the absence or the want of all discovery, the finger of a God is disowned; and it seems to mark how resolute and at the same time how hopeless is the infidelity of modern times, that, just in proportion to our ignorance of all the secondary or the sensible causes, is our haughty refusal of any homage to the first cause. It is passing strange of this disease, that, after having baffled every attempt to find out its dependence on aught that is on earth, the idea of its dependence on the will of Heaven should of all others have been laughed most impiously to scorn. The voice of derision and defiance was first heard in our high places; and thence it passed, as if by infection, into general society. And so, many have disowned the power and the will of the Deity in this visitation. They most unphilosophically, we think, as well as impiously have spurned at prayer.

But we cannot pass away from this part of our subject, without adverting to a recent event, the thought of which is at present irresistibly obtruded on us, and by which this parish and congregation but a few weeks ago have been deprived of one of the most conspicuous of our office-bearers—one who constitutionally the kindest and most indulgent of men, was the most alive of all I ever knew to the wants and the miseries of our common nature; and who finely alive to all the impulses and soft touches of humanity, laboured night and day in the vocation of doing good continually.

But, instead of saying that he laboured, I should say that he luxuriated in well doing; for never was a heart more attuned to ready and responsive agreement with the calls of benevolence than his, and sooner would I believe of nature that she had receded from her constancy, than of him that e'er

> "He looked unmoved on misery's languid eye,
> Or heard her sinking voice without a sigh."

Of all the recollections which the friends either of my youth or of my manhood have left behind them in this land of dying men, there is none more beautifully irradiated—whether I look back on the mildness of his christian worth, or on those sensibilities of an open and generous and finely attempered spirit, which gives such a charm to human companionship. And as the second great law is like unto the first; so that love of his which went forth so diffusively amongst his fellows upon earth, we humbly hope, was at once the indication and the consequent of a love that ascended with high and habitual aspiration to God in heaven. It was through a brief and tremendous agony that he was carried from the world of sense to the world of spirits; and yet it is a happiness to be told that the faith and hope of the Gospel lighted up a halo over his expiring moments, and that, ere death had closed his eyes, he through nearly an hour of audible prayer gave his last testimony to the truth as it is in Jesus.*

* This notice refers to John Wilson, Esq., Silk merchant in Glasgow, who was Kirk Treasurer of St. John's, and to the deep regret of all who knew him, was carried off by cholera in the neighbourhood of Glasgow.

But to recal ourselves from this theme of sadness, we trust you will now understand of every event in nature or history, that each in the order of causation is preceded by a train which went before it, and that man's observations can extend more or less a certain way along this train, till they are lost in the undiscovered and at length undiscoverable recesses which are placed beyond the cognizance of the human faculties. Now it is because of the higher and unknown part which belongs to every such series that we bid you respect the lessons of piety, for God hath not so constructed the universe as to remove it from the hold of His own special management and superintendence; and therefore, not in one thing the Bible tells us, but in every thing we should make our requests known unto God. But again, it is because of the lower and the known or ascertained and strictly uniform part which belongs to every series, that we bid you respect the lessons of experience; for God did not so conduct the affairs of His universe, as to thrust forth His invisible hand among its visible successions; but while He keeps a perpetual and ascendant hold among the springs of that machinery which is behind the curtain, He leaves untouched all those wonted regularities, which, on the stage of observation, are patent to human eyes. Now these are the respective domains of philosophy and faith, and this is the use to be made of them. Looking to the one, we learn the subordination of all nature. Looking to the other, we learn the constancy of visible nature. These great truths harmonize;

and between the lessons which they give, there is the fullest harmony. He who is enlightened and acts upon both is at one and the same time a man of prudence and a man of prayer; who never loses his confidence in God, yet, as awake to the manifestations of experience as if they were the manifestations of the divine will, never counts upon a miracle. He holds perpetual converse with heaven; yet shapes his earthly conduct by his earthly circumstances. In his habits of diligence he proceeds on the uniformity of visible nature, and he does accordingly. In his habits of devotion he knows that there is a visible power above which subordinates all nature, and he prays accordingly. He is neither the mystic who will not act, nor is he the infidel who will not pray. He knows how to combine both, or how to combine wisdom with piety—that rare and beauteous combination unknown to the world at large, yet realized by many a cottage patriarch, who, without attempting, without being capable in fact of any profound or philosophical adjustment between them, but on his simple understanding alone of Scripture lessons and Scripture examples, unites the most strenuous diligence in the use of means, with the strictest dependence upon God. Without the combination of these two, there has been nothing great, nothing effective in the history of the church; and, on the other hand, we find that all the most illustrious, whether in philanthropy or in christian patriotism, from the apostle Paul to the highest names in the descending history of the world, as Augustine and Luther and Knox and Howard, that, superadding the wisdom of experience to a sense of deepest piety, they were at once men of performance and men of prayer.

But let us look for a moment to the highest example of all, even that of our Saviour when on earth; for in the history of His temptation, will the eye of the diligent observer recognize an application and a moral, which serve, we think very finely, to illustrate our whole argument.

The first proposal of the adversary was, that, because an hungered by the abstinence of forty days and forty nights in the wilderness, he should turn stones into bread; and the reply of our Saviour that "man liveth not by bread alone but by every word which cometh out of the mouth of God" bespoke His confidence in that Supreme Power which overrules all nature. Now observe how this is followed up by the tempter—since such His confidence I may perhaps prevail upon Him to cast Himself from the pinnacle of the temple, employing the very argument He just has used, even the overruling power of that God who can bear Him up by the intervention of angels lest He dash his foot against a stone. The reply "thou shalt not tempt the Lord thy God" tells us, that the same Being who overrules all nature, never interferes, but for some worthy and great purpose to thwart the established successions of visible nature; and that it is wrong, it is wanton, in any of His creatures so to act, as if he counted upon such an interference. It is a noble lesson for us never to traverse or neglect the means which experience hath told us are effectual for good; and never to brave, but at the call of imperious duty, the exposures which the same experience has told us, on our knowledge or recollection of Nature's established processes, are followed up by evil. Our Saviour would not, in defiance to the law of gravitation, cast Himself off from that place of security which upheld Him against its power. And neither should we ever, though in defiance but to the probable law of contagion, or by what (to borrow a usual phrase) might well be termed a tempting of Providence, refuse those places or cast away those measures of security, that are found to protect us against the virulence of this destroyer. In a word, between the wisdom of piety and the wisdom of experience there is most profound harmony—unknown to the infidel, and so he hath cast off prayer; unknown to the fanatic, and so he hath cast prudence away from him.

And we appeal to you, my brethren, if there be not much in the state and recent history of our nation to confirm these views. We rejoiced in the appointment several months ago of a national fast, and that notwithstanding the contempt and annoyance of the many infidel manifestations to which the appointment had been exposed—hoping, as we then did, that it would meet with a duteous and a general response from the people of the land; and perceiving afterwards, in our limited sphere, the obvious solemnity, and we trust in a goodly number of instances, the deep and heart-felt sacredness of its observation among our families. It is well that there should be a public and a prayerful recognition of God in the midst of us; and we have failed in our argument, we have failed, whether from the obscurity of its illustrations or the obscurity of its terms, in obtaining for it the sympathy of your understandings—if you perceive not, that, in the distinct relation of cause and effect, there is a real substantive connection between the supplications which ascend for health and safety from the midst of a land, and the actual warding off of disease and death from its habitations. But in fullest harmony with this it is also well, I would go farther and say there is no infringement upon deepest piety in pronouncing it indispensable—that while we invoke the Heavenly Agent who sitteth above for every effectual blessing, all the earthly means and earthly instruments should be in complete and orderly preparation. We are aware that in many places and on many occasions these have been rebelled against.* And it but enhances the lesson, beside carrying a most impressive rebuke, both to the fanaticism of an ill-understood Christianity; and to the ignorant frenzy of an ill-educated and, in respect to the woeful deficiency both of churches and schools, we would say a neglected population—that just in those places where the offered help of the physician was most strenuously and most

* In Edinburgh the metropolis of medical science, a vigorous system of expedients was instituted; and nothing could exceed the promptitude and the watchfulness and the activity, at a moment's call, wherewith the disease was met and repressed at every point of its outbreakings. And we cannot imagine a more striking demonstration for the importance of human agency, diligently operating on all the resources which Nature and experience have placed within our reach, than is furnished by a comparison between the perfection of our city arrangements, and the fewness of our city deaths.

ungratefully resisted, and at times indeed by violence overborne, that there it was where the disease reasserted its power, and as if with the hand of an avenger shook menace and terror among the families. As if the same God who bids us in His word make request unto Him in all things, would furthermore tell us by His Providence, that, in no one thing will He permit a heedless invasion on the regularities of that course which He Himself has established; that with His own hand He ordained the footsteps of Nature, and He will chastise the presumption of those who shall think to contravene the ordinance; that experience is the school-master authorized by Him for the government and guidance of His family on earth, and that He will resent the outrage done to her authority whenever her lessons or her laws are wantonly violated.

In conclusion let us observe, that, on the one hand, we shall be glad if aught that has been said will help to conciliate our mere religionists to the lessons of experience and of sound philosophy; and, in opposition to those senseless prejudices, by which they have often brought the most unmerited derision and discredit on their own cause, we would remind them that it is not all philosophy which Scripture denounces, but only vain philosophy—it is not all science which it deprecates, but only the science falsely so called. On the other hand we should rejoice in witnessing the mere philosopher, or man of secular and experimental wisdom, more conciliated than he is to the lessons of Religion, and to that humble faith which is the great and actuating spirit of its observations and its pieties and its prayers. We have heard that the study of Natural Science disposes to Infidelity. But we feel persuaded that this is a danger only associated with a slight and partial, never with a deep and adequate and comprehensive view of its principles. It is very possible that the conjunction between science and scepticism may at present be more frequently realised than in former days; but this is only because, in spite of all that is alleged about this our more enlightened day and more enlightened public, our science is neither so deeply founded nor of such firm and thorough staple as it wont to be. We have lost in depth what we have gained in diffusion—having neither the massive erudition, nor the gigantic scholarship, nor the profound and well-laid philosophy of a period that has now gone by; and it is to this that infidelity stands indebted for her triumphs among the scoffers and the superficialists of a half-learned generation.

DISCOURSE III.

THE TRANSITORY NATURE OF VISIBLE THINGS.

"The things which are seen are temporal."—2 Cor. iv. 18.

The assertion that the things which are seen are temporal, holds true in the absolute and universal sense of it. They had a beginning, and they will have an end. Should we go upward through the stream of ages that are past, we come to a time when they were not. Should we go onward through the stream of ages that are before us, we come to a time when they will be no more. It is indeed a most mysterious flight which the imagination ventures upon, when it goes back to the eternity that is behind us—when it mounts its ascending way through the millions and the millions of years that are already gone through, and stop where it may, it finds the line of its march always lengthening beyond it, and losing itself in the obscurity of as far removed a distance as ever. It soon reaches the commencement of visible things, or that point in its progress when God made the heavens and the earth. They had a beginning, but God had none; and what a wonderful field for the fancy to expatiate on, when we get above the era of created worlds, and think of that period when, in respect of all that is visible, the immensity around us was one vast and unpeopled solitude. But God was there in his dwelling place, for it is said of Him, that He inhabits eternity; and the Son of God was there, for we read of the glory which He had with the Father before the world was. The mind cannot sustain itself under the burden of these lofty contemplations. It cannot lift the curtain which shrouds the past eternity of God. But it is good for the soul to be humbled under a sense of its incapacity. It is good to realize the impression which too often abandons us, that He made us, and not we ourselves. It is good to feel how all that is temporal lies in passive and prostrate subordination before the will of tne uncreated God. It is good to know how little a portion it is that we see of Him and of His mysterious ways. It is good to lie at the feet of His awful and unknown majesty—and while secret things belong to Him, it is good to bring with us all the helplessness and docility of children to those revealed lessons which belong to us and to our children.

But this is not the sense in which the temporal nature of visible things is taken up by the Apostle. It is not that there is a time past in which they did not exist—but that there is a time to come in which they will exist no more. He calls them temporal, because the time and the duration of their existence will have an end. His eye is full upon futurity. It is the passing away of visible things in the time that is to come, and the ever during nature of invisible things through the eternity that is to come, which the Apostle is contemplating. Now, on this one point we say nothing about the positive annihilation of the matter of visible things. There is reason for believing, that some of the matter of our present bodies may exist in those more glorified and transformed bodies which we are afterwards to occupy. And for any thing we know, the matter of the present world, and of the present system may exist in those new heavens and that new earth, wherein dwelleth righteousness. There may be a transfiguration of matter without a destruction of it —and, therefore it is, that when we assert with the Apostle in the text, how things seen are temporal, we shall not say more than that the substance of these things, if not consigned back again to the nothing from which they had emerged, will be employed in the formation of other things totally different—that the change will be so great, as that all old things may be said to have passed away, and all things to become new—that after the wreck of the last conflagration, the desolated scene will be repeopled with other objects; the righteous will live in another world, and the eye of the glorified body will open on another field of contemplation from that which is now visible around us.

Now, in this sense of the word temporal, the assertion of my text may be carried round to all that is visible. Even those objects which men are most apt to count upon as unperishable, because, without any sensible decay, they have stood the lapse of many ages, will not weather the lapse of eternity. This earth will be burnt up. The light of yonder sun will be extinguished. These stars will cease from their twinkling. The heavens will pass away as a scroll—and as to those solid and enormous masses which, like the firm world we tread upon, roll in mighty circuit through the immensity around us, it seems the solemn language of revelation of one and all of them, that from the face of Him who sitteth on the throne, the earth and the heavens will fly away, and there will be found no place for them.

Even apart from the Bible, the eye of observation can witness, in some of the hardest and firmest materials of the present system, the evidence of its approaching dissolution. What more striking, for example, than the natural changes which take place on the surface of the world, and which prove that the strongest of Nature's elements must, at last yield to the operation of time and of decay—that yonder towering mountain, though propped by the rocky battlements which surround it, must at last sink under the power of corruption—that every year brings it nearer to its end—that at this moment, it is wasting silently away, and letting itself down from the lofty eminence which it now occupies—that the torrent which falls from its side never ceases to consume its substance, and to carry it off in the form of sediment to the ocean—that the frost which assails it in winter loosens the solid rock, detaches it in pieces from the main precipice, and makes it fall in fragments to its base—that the power of the weather scales off the most flinty materials, and that the wind of heaven scatters them in dust over the surrounding country—that even though not anticipated by the sudden and awful convulsions of the day of God's wrath, nature contains within itself the rudiments of decay—that every hill must be levelled with the plains, and every plain be swept away by the constant operation of the rivers which run through it—and that, unless renewed by the hand of the Almighty, the earth on which we are now treading must disappear in the mighty roll of ages and of centuries. We cannot take our flight to other worlds, or have a near view of the changes to which they are liable. But surely if this world, which, with its mighty apparatus of continents and islands, looks so healthful and so firm after the wear of many centuries, is posting visibly to its end, we may be prepared to believe that the principles of destruction are also at work in other provinces of the visible creation—and that though of old God laid the foundation of the earth, and the heavens are the work of his hands, yet they shall perish; yea, all of them shall wax old like a garment, and as a vesture shall He change them, and they shall be changed.

We should be out of place in all this style of observation, did we not follow it up with the sentiment of the Psalmist, "These shall perish, but thou shalt endure; for thou art the same, and thy years have no end." What a lofty conception does it give us of the majesty of God, when we think how He sits above, and presides in high authority over this mighty series of changes—when after sinking under our attempts to trace him through the eternity that is behind, we look on the present system of things, and are taught to believe that it is but a single step in the march of His grand administrations through the eternity that is before us—when we think of this goodly universe, summoned into being to serve some temporary evolution of His great and mysterious plan—when we think of the time when it shall be broken up, and out of its disordered fragments other scenes and other systems shall emerge—surely, when fatigued with the vastness of these contemplations, it well becomes us to do the homage of our reverence and wonder to the one Spirit which conceives and animates the whole, and to the one noble design which runs through all its fluctuations.

But there is another way in which the objects that are seen are temporal. The object may not merely be removed from us, but we may be removed from the object. The disappearance of this earth, and of these heavens from us, we look upon through the dimness of a far-placed futurity. It is an event, therefore, which may regale our imagination; which may lift our mind by its sublimity; which may disengage us in the calm hour of meditation from the littleness of life, and of its cares; and which may even throw a clearness and a solemnity over our intercourse with God. But such an event as this does not come home upon our hearts with the urgency of a personal interest. It does not carry along with it the excitement which lies in the nearness of an immediate concern. It does not fall with such vivacity upon our conceptions, as practically to tell on our pursuits, or any of our purposes. It may elevate and solemnize us, but this effect is perfectly consistent with its having as little influence on the walk of the living, and the moving, and the acting man, as a dream of poetry. The Preacher may think that he has done great things with his eloquence—and the hearers may think that great things have been done upon them—for they felt a fine glow of emotion, when they heard of God sitting in the majesty of His high counsels, over the progress and the destiny of created things. But the truth is, that all this kindling of devotion which is felt upon the contemplation of His greatness, may exist in the same bosom, with an utter distaste for the holiness of His character; with an entire alienation of the heart and of the habits from the obedience of His law; and above all, with a most nauseous and invincible contempt for the spiritualities of that revelation, in which He has actually made known His will and His ways to us. The devotion of mere taste is one thing—the devotion of principle is another. And as surely as a man may weep over the elegant sufferings of poetry, yet add to the real sufferings of life by peevishness in his family, and insolence among his neighbours—so, surely may a man be wakened to rapture by the magnificence of God, while his life is deformed by its rebellions, and his heart rankles with all the foulness of idolatry against Him.

Well, then, let us try the other way of bringing the temporal nature of visible things to bear upon your interests. It is true, that this earth, and these heavens, will at length disappear; but they may outlive our posterity for many generations. However, if they disappear not from us, we most certainly shall disappear from them. They will soon cease to be any thing to you—and though the splendour and variety of all that is visible around us, should last for thousands of centuries, your eyes will soon be closed upon them. The time is coming when this goodly scene shall reach its positive consummation. But, in all likelihood, the time is coming much sooner, when you shall resign the breath of your nostrils, and bid a final adieu to every thing around you. Let this earth, and these heavens be as enduring as they may, to you they are fugitive as vanity. Time, with its mighty strides, will soon reach a future generation, and leave the present in death and in forgetfulness behind it. The grave will close upon every one of you, and that is the dark and the silent cavern where no voice is heard, and the light of the sun never enters.

But more than this. Though we live too short a time to see the great changes which are carrying on in the universe, we live long enough to see many of its changes—and such changes too as are best fitted to warn and to teach us; even the changes which take place in society, made up of human beings as frail and as fugitive as ourselves. Death moves us away from many of those objects which are seen and temporal—but we live long enough to see many of these objects moved away from us—to see acquaintances falling every year—to see families broken up by the rough and unsparing hand of death—to see houses and neighbourhoods shifting their inhabitants—to see a new race, and a new generation—and, whether in church or in market, to see unceasing changes in the faces of the people who repair to them. We know well, that there is a poetic melancholy inspired by such a picture as this, which is altogether unfruitful—and that totally apart from religion, a man may give way to the luxury of tears, when he thinks how friends drop away from him—how every year brings along with it some sad addition to the registers of death—how the kind and hospitable mansion is left without a tenant—and how, when you knock at a neighbour's door, you find that he who welcomed you, and made you happy, is no longer there. O that we could impress by all this, a salutary direction on the fears and on the consciences of individuals—that we could give them a living impression of that coming day, when they shall severally share in the general wreck of the species—when each of you shall be one of the many whom the men of the next generation may remember to have lived in yonder street, or laboured in yonder manufactory—when they shall speak of you, just as you speak of the men of the former generation—who, when they died, had a few tears dropped over their memory, and for a few years will still continue to be talked of. O could we succeed in giving you a real and living impression of all this; and then may we hope to carry the lesson of John the Baptist with energy to your fears, "Flee from the coming wrath." But there is something so very deceiving in the progress of time. Its progress is so gradual. To-day is so like yesterday that we are not sensible of its departure. We should make head against this delusion. We should turn to personal account every example of change or of mortality. When the clock strikes, it should remind you of the dying hour. When you hear the sound of the funeral bell, you should think, that in a little time it will perform for you the same office. When you wake in the morning, you should think that there has been the addition of another day to the life that is past, and the subtraction of another day from the remainder of your journey. When the shades of the evening fall around you, you should think of the steady and invariable progress of time—how the sun moves and moves till it will see you out—and how it will continue to move after you die, and see out your children's children to the latest generations. Every thing around us should impress the mutability of human affairs. An acquaintance dies—you will soon follow him. A family moves from the neighbourhood—learn that the works of man are given to change. New families succeed—sit loose to the world, and withdraw your affections from its unstable and fluctuating interests. Time is rapid, though we observe not its rapidity. The days that are past appear like the twinkling of a vision. The days that are to come will soon have a period, and will appear to have performed their course with equal rapidity. We talk of our fathers and our grandfathers, who figured their day in the theatre of the world. In a little time, we will be the ancestors of a future age. Posterity will talk of us as of the men that are gone—and our remembrance will soon depart from the face of the country. When we attend the burial of an acquaintance, we see the bones of the men of other times—in a few years, our bodies will be mangled by the power of corruption, and be thrown up in loose and scattered fragments among the earth of the new made grave. When we wander among the tombstones of the church-yard we can scarcely follow the mutilated letters that compose the simple story of the inhabitant below. In a little time, and the tomb that covers us will moulder by the power of the seasons —and the letters will be eaten away—and the story that was to perpetuate our remembrance, will elude the gaze of some future inquirer.

We know that time is short, but none of us know how short. We know that it will not go beyond a certain limit of years; but none of us know how small the number of years, or months, or days may be. For death is at work upon all ages. The fever of a few days may hurry the likeliest of us all from this land of mortality. The cold of a few weeks may settle into some lingering but irrecoverable disease. In one instant the blood of him who has the promise of many years, may cease its circulation. Accident may assail us. A slight fall may precipitate us into eternity. An exposure to rain may lay us on the bed of our last sickness, from which we are never more to rise. A little spark may kindle the midnight conflagration, which lays a house and its inhabitants in ashes. A stroke of lightning may arrest the current of life in a twinkling. A gust of wind may overturn the vessel, and lay the unwary passenger in a watery grave. A thousa d dangers beset us on the slippery path of this world; and no age is exempted from them —and from the infant that hangs on its mother's bosom, to the old man who sinks under the decrepitude of years, we see death in all its woeful and affecting varieties.

You may think it strange—but even still we fear,

we may have done little in the way of sending a fruitful impression into your consciences. We are too well aware of the distinction between seriousness of feeling, and seriousness of principle, to think that upon the strength of any such moving representation as we are now indulging in, we shall be able to dissipate that confounded spell which chains you to the world, to reclaim your wandering affections, or to send you back to your week-day business more pure, and more heavenly. But sure we are you ought to be convinced, that all which binds you so cleavingly to the dust is infatuation and vanity; that there is something most lamentably wrong in your being carried away by the delusions of time—and this is a conviction which should make you feel restless and dissatisfied. We are well aware, that it is not human eloquence, or human illustration, that can accomplish a victory over the obstinate principles of human corruption—and therefore it is that we feel as if we did not advance aright through a single step of a sermon, unless we look for the influences of that mighty Spirit, who alone is able to enlighten and arrest you—and may employ even so humble an instrument as the voice of a fellow mortal, to send into your heart the inspiration of understanding.

We now shortly insist on the truth, that the things which are not seen are eternal. No man hath seen God at any time, and He is eternal. It is said of Christ, " whom having not seen, we love, and he is the same to-day, yesterday, and for ever." It is said of the Spirit, that, like the wind of heaven, He eludes tne observation, and no man can tell of him whence He cometh, or whither He goeth—and He is called the Eternal Spirit, through whom the Son offered Himself up without spot unto God. We are quite aware, that the idea suggested by the eternal things which are spoken of in our text, is heaven, with all its circumstances of splendour and enjoyment. This is an object which, even on the principles of taste, we take a delight in contemplating: and it is also an object set before us in the Scriptures, though with a very sparing and reserved hand. All the descriptions we have of heaven there, are general, very general. We read of the beauty of the heavenly crown, of the unfading nature of the heavenly inheritance, of the splendour of the heavenly city—and these have been seized upon by men of imagination, who, in the construction of their fancied paradise, have embellished it with every image of peace, and bliss, and loveliness; and, at all events, have thrown over it that most kindling of all conceptions, the magnificence of eternity. Now, such a picture as this has the certain effect of ministering delight to every glowing and susceptible imagination. And here lies the deep-laid delusion, which we have occasionally hinted at. A man listens, in the first instance, to a pathetic and high wrought narrative on the vanities of time—and it touches him even to the tenderness of tears. He looks, in the second instance, to the fascinating perspective of another scene, rising in all the glories of immortality from the dark ruins of the tomb, and he feels within him all those ravishments of fancy, which any vision of united grandeur and loveliness would inspire. Take these two together, and you have a man weeping over the transient vanities of an ever-shifting world, and mixing with all this softness, an elevation of thought and of prospect, as he looks through the vista of a futurity, losing itself in the mighty range of thousands and thousands of centuries. And at this point the delusion comes in, that here is a man who is all that religion would have him to be—a man weaned from the littleness of the paltry scene that is around him—soaring high above all the evanescence of things present, and things sensible—and transferring every affection of his soul to the durabilities of a pure and immortal region. It were better if this high state of occasional impression on the matters of time and of eternity, had only the effect of imposing the falsehood on others, that the man who was so touched and so transported, had on that single account the temper of a candidate for heaven. But the falsehood takes possession of his own heart. The man is pleased with his emotions and his tears—and the interpretation he puts upon them is, that they come out of the fulness of a heart all alive to religion, and sensibly affected with its charms, and its seriousness, and its principle. Now, we venture to say, that there may be much of all this kind of enthusiasm, with the very man who is not moving a single step towards that blessed eternity, over which his fancy delights to expatiate. The moving representation of the preacher may be listened to as a pleasant song—and the entertained hearer return to all the inveterate habits of one of the children of this world.

It is this, which makes us fear that a power of deceitfulness may accompany the eloquence of the pulpit—that the wisdom of words may defeat the great object of a practical work upon the conscience —that a something short of a real business change in the heart, and in the principles of acting, may satisfy the man who listens, and admires, and resigns his every feeling to the magic of an impressive description—that, strangely compounded beings as we are, broken loose from God, and proving it by the habitual voidness of our hearts to a sense of His authority, and of His will; that blind to the realities of another world, and slaves to the wretched infatuation which makes us cleave with the full bent of our affections to the one by which we are visibly and immediately surrounded; that utterly unable, by nature, to live above the present scene, while its cares, and its interests are plying us every hour with their urgency; that the prey of evil passions which darken and distract the inner man, and throw us at a wider distance from the holy Being who forbids the indulgence of them; and yet with all this weight of corruption about us, having minds that can seize the vastness of some great conception, and can therefore rejoice in the expanding loftiness of its own thoughts, as it dwells on the wonders of eternity; and having hearts that can move to the impulse of a tender consideration, and can, therefore, sadden into melancholy at the dark picture of death, and its unrelenting cruelties; and having fancies that can brighten to the cheerful colouring of some pleasing and hopeful representation, and can, therefore, be soothed and animated when some sketch is laid before it of a pious family emerging from a common sepulchre, and on the morning of their joyful resurrection, forgetting all the sorrows and separations of the dark world that has now rolled over them—O my brethren, we fear it, we greatly fear it, that while busied with topics such as these, many a hearer may weep, or be elevated, or take pleasure in the touching imagery that is made to play around him, while the dust of this perishable earth is all that his soul cleaves to—and its cheating vanities are all that his heart cares for, or his footsteps follow after.

The thing is not merely possible—but we see in it a stamp of likelihood to all that experience tells us of the nature or the habitudes of man. Is there no such thing as his having a taste for the beauties of landscape, and, at the same time, turning with disgust from what he calls the methodism of peculiar Christianity? Might not he be an admirer of poetry, and, at the same time, nauseate with his whole heart, the doctrine and the language of the New Testament? Might not he have a fancy that can be regaled by some fair and well-formed vision of immortality—and, at the same time, have no practical hardihood whatever for the exercise of labouring in the prescribed way after the meat that endureth? Surely, surely, this is all very possible—and it is just as possible, and many we believe to be the instances we have of it in real life, when an eloquent description of heaven is exquisitely felt, and wakens in the bosom the raptures of the sincerest admiration, among those who feel an utter repugnancy to the heaven of the Bible—and are not moving a single inch through the narrowness of the path which leads to it.

DISCOURSE IV.

ON THE NEW HEAVENS AND THE NEW EARTH.

"Nevertheless we, according to his promise, look for new heavens and a new earth, wherein dwelleth righteousness."—2 PETER iii. 13.

THERE is a limit to the revelations of the Bible about futurity, and it were a mental or spiritual trespass to go beyond it. The reserve which it maintains in its informations, we also ought to maintain in our inquiries—satisfied to know little on every subject, where it has communicated little, and feeling our way into regions which are at present unseen, no further than the light of Scripture will carry us.

But while we attempt not to be "wise above that which is written," we should attempt, and that most studiously, to be wise up to that which is written. The disclosures are very few and very partial, which are given to us of that bright and beautiful economy, which is to survive the ruins of our present one. But, still there are such disclosures—and on the principle of the things that are revealed belonging unto us, we have a right to walk up and down, for the purpose of observation, over the whole actual extent of them. What is made known of the details of immortality, is but small in the amount, nor are we furnished with the materials of any thing like a graphical or picturesque exhibition of its abodes of blessedness. But still somewhat is made known, and which, too, may be addressed to a higher principle than curiosity, being like every other Scripture, " profitable both for doctrine and for instruction in righteousness."

In the text before us, there are two leading points of information, which we should like successively to remark upon. The first is, that in the new economy which is to be reared for the accommodation of the blessed, there will be materialism, not merely new heavens, but also a new earth. The second is, that as distinguished from the present, which is an abode of rebellion, it will be an abode of righteousness.

I. We know historically that earth, that a solid material earth, may form the dwelling of sinless creatures, in full converse and friendship with the Being who made them—that, instead of a place of exile for outcasts, it may have a broad avenue of communication with the spiritual world, for the descent of ethereal beings from on high—that, like the member of an extended family, it may share in the regard and attention of the other members, and along with them be gladdened by the presence of Him who is the Father of them all. To inquire how this can be, were to attempt a wisdom beyond Scripture: but to assert that this has been, and therefore may be, is to keep most strictly and modestly within the limits of the record. For, we there read, that God framed an apparatus of materialism, which, on His own surveying, He pronounced to be all very good, and the leading features of which may still be recognized among the things and the substances that are around us—and that He created man with the bodily organs and senses which we now wear—and placed Him under the very canopy that is over our heads—and spread around Him a scenery, perhaps lovelier in its tints, and more smiling and serene in the whole aspect of it, but certainly made up, in the main, of the same objects that still compose the prospect of our visible contemplations—and there, working with his hands in a garden, and with trees on every side of him, and even with animals sporting at his feet, was this inhabitant of earth, in the midst of all those earthly and familiar accompaniments, in full possession of the best immunities of a citizen of heaven—sharing in the delight of angels, and while he gazed on the very beauties which we ourselves gaze upon, rejoicing in them most as the tokens of a present and presiding Deity. It were venturing on the region of conjecture to affirm, whether, if Adam had not fallen, the earth that we now tread upon, would have been the everlasting abode of him and his posterity. But certain it is, that man, at the first, had for his place this world, and, at the same time, for his privilege, an unclouded fellowship with God, and, for his prospect, an immortality, which death was neither to intercept nor put an end to. He was terrestrial in respect of condition, and yet celestial in respect both of character and enjoyment. His eye looked outwardly on a landscape of earth, while his heart breathed upwardly in the love of heaven. And though he trode the solid platform of our world, and was compassed about with its horizon—still was he within the circle of God's favoured creation, and took His place among the freemen and the denizens of the great spiritual commonwealth.

This may serve to rectify an imagination, of which we think that all must be conscious—as if the grossness of materialism was only for those who had degenerated into the grossness of sin; and that, when a spiritualizing process had purged away all our corruption, then, by the stepping stones of a death and a resurrection, we should be borne away to some ethereal region, where sense, and body, and all in the shape either of audible sound, or of tangible substance, were unknown. And hence that strangeness of impression which is felt by you, should the supposition be offered, that in the place of eternal blessedness, there will be ground to walk upon; or scenes of luxuriance to delight the corporeal senses; or the kindly intercourse of friends talking familiarly, and by articulate converse together; or, in short, any thing that has the least resemblance to a local territory, filled with various accommodations, and peopled over its whole extent by creatures formed like ourselves—having bodies such as we now wear, and faculties of perception, and thought, and mutual communication, such as we now exercise. The common imagination that we have of paradise on the other side of death, is, that of a lofty aërial region, where the inmates float in ether, or are mysteriously suspended upon nothing —where all the warm and sensible accompaniments which give such an expression of strength, and life, and colouring, to our present habitation, are attenuated into a sort of spiritual element, that is meagre, and imperceptible, and utterly uninviting to the eye of mortals here below—where every vestige of materialism is done away, and nothing left but certain unearthly scenes that have no power of allurement, and certain unearthly ecstasies, with which it is felt impossible to sympathise. The holders of this imagination forget all the while, that really there is no essential connexion between materialism and sin—that the world which we now inhabit, had all the amplitude and solidity of its present materialism, before sin entered into it—that God so far, on that account, from looking slightly upon it, after it had received the last touch of His creating hand, reviewed the earth, and the waters, and the firmament, and all the green herbage, with the living creatures, and the man whom He had raised in dominion over them, and He saw every thing that He had made, and behold it was all very good. They forget that on the birth of materialism, when it stood out in the freshness of those glories which the great Architect of Nature had impressed upon it, that then "the morning stars sang together, and all the sons of God shouted for joy." They forget the appeals that are made everywhere in the Bible to this material workmanship—and how, from the face of these visible heavens, and the garniture of this earth that we tread upon, the greatness and the goodness of God are reflected on the view of His worshippers. No, my brethren, the object of the administration we sit under, is to extirpate sin, but it is not to sweep away materialism. By the convulsions of the last day, it may be shaken, and broken down from its present arrangements; and thrown into such fitful agitations, as that the whole of its existing framework shall fall to pieces; and with a heat so fervent as to melt its most solid elements, may it be utterly dissolved. And thus may the earth again become without form, and void, but without one particle of its substance going into annihilation. Out of the ruins of this second chaos, may another heaven and another earth be made to arise; and a new materialism, with other aspects of magnificence and beauty, emerge from the wreck of this mighty transformation; and the world be peopled as before, with the varieties of material loveliness, and space be again lighted up into a firmament of material splendour.

Were our place of everlasting blessedness so purely spiritual as it is commonly imagined, then the soul of man, after, at death, having quitted his body, would quit it conclusively. That mass of materialism with which it is associated upon earth, and which many regard as a load and an incumbrance, would have leave to putrefy in the grave, without being revisited by supernatural power, or raised again out of the inanimate dust into which it had resolved. If the body be indeed a clog and a confinement to the spirit, instead of its commodious tenement, then would the spirit feel lightened by the departure it had made, and expatiate in all the buoyancy of its emancipated powers, over a scene of enlargement. And this is, doubtless, the prevailing imagination. But why then, after having made its escape from such a thraldom, should it ever recur to the prison-house of its old materialism if a prison-house it really be. Why should the disengaged spirit again be fastened to the drag of that grosser and heavier substance, which many think has only the effect of weighing down its activity, and infusing into the pure element of mind an ingredient which serves to cloud and to enfeeble it. In other words, what is the use of a day of resurrection, if the union which then takes place is to deaden, or to reduce all those energies that are commonly ascribed to the living principle, in a state of separation? But, as a proof of some metaphysical delusion upon this subject, the product, perhaps, of a wrong though fashionable philosophy, it would appear, that to embody the spirit is not the stepping-stone to its degradation, but to its preferment. The last day will be a day of triumph to the righteous—because the day of the re-entrance of the spirit to its much-loved abode, where its faculties, so far from being shut up into captivity, will find their free and kindred development in such material organs as are suited to them. The fact of the resurrection proves, that, with man at least, the state of a disembodied spirit, is a state of unnatural violence—and that the resurrection of his body is an essential step to the highest perfection of which he is susceptible. And it is indeed an homage to that materialism, which many are for expunging from the future state of the universe altogether—that ere the immaterial soul of man has reached the ultimate glory and blessedness which are designed for it, it must return and knock at that very grave where lie the mouldered remains of the body which it wore—and there inquisition must be made for the flesh, and the sinews, and the bones, which the power of corruption has perhaps for centuries before, assimilated to the earth that is around them—and there, the minute atoms must be re-assembled into a structure that bears upon it the form and the lineaments, and the general aspect of a man—and the soul passes into this material framework, which is hereafter to be its lodging-place for ever—and that, not as its prison, but as its pleasant and befitting habitation—not to be trammelled, as some would have it, in a hold of materialism, but to be therein equipped for the services of eternity—to walk embodied among the bowers of our second paradise—to stand embodied in the presence of our God.

There will, it is true, be a change of personal constitution between a good man before his death, and a good man after his resurrection—not, however, that he will be set free from his body, but that he will be set free from the corrupt principle which is in his body—not that the materialism by which he is now surrounded will be done away, but that the taint of evil by which this materialism is now pervaded, will be done away. Could this be effected without dying, then death would be no longer an essential stepping-stone to paradise. But it would appear of the moral virus which has been transmitted downwards from Adam, and is now spread abroad over the whole human family—it would appear, that to get rid of this, the old fabric must be taken down, and reared anew; and that, not of other materials, but of its own materials, only delivered of all impurity, as if by a refining process in the sepulchre. It is thus, that what is "sown in weakness, is raised in power"—and for this purpose, it is not necessary to get quit of materialism, but to get quit of sin, and so to purge materialism of its malady. It is thus that the dead shall come forth incorruptible—and those, we are told, who are alive at this great catastrophe, shall suddenly and mysteriously be changed. While we are compassed about with these vile bodies, as the apostle emphatically terms them, evil is present, and it is well, if through the working of the Spirit of grace, evil does not prevail. To keep this besetting enemy in check, is the task and the trial of our Christianity on earth—and it is the detaching of this poisonous ingredient which constitutes that for which the believer is represented as groaning earnestly, even the redemption of the body that he now wears, and which will then be transformed into the likeness of Christ's glorified body. And this will be his heaven, that he will serve God without a struggle, and in a full gale of spiritual delight—because with the full concurrence of all the feelings and all the faculties of his regenerated nature. Before death, sin is only repressed—after the resurrection, sin will be exterminated. Here he has to maintain the combat, with a tendency to evil still lodging in his heart, and working a perverse movement among his inclinations; but after his warfare in this world is accomplished, he will no longer be so thwarted—and he will set him down in another world, with the repose and the triumph of victory for his everlasting reward. The great constitutional plague of his nature will no longer trouble him; and there will be the charm of a general affinity between the purity of his heart, and the purity of the element he breathes in. Still it will not be the purity of spirit escaped from materialism, but of spirit translated into a materialism that has been clarified of evil. It will not be the purity of souls unclothed as at death, but the purity of souls that have again been clothed upon at the resurrection.

But the highest homage that we know of to materialism, is that which God, manifest in the flesh, has rendered to it. That He, the Divinity, should have wrapt His unfathomable essence in one of its coverings, and expatiated amongst us in the palpable form and structure of a man; and that He should have chosen such a tenement, not as a temporary abode, but should have borne it with Him to the place which He now occupies, and where He is now employed in preparing the mansions of His followers—that He should have entered within the vail, and be now seated at the right hand of the Father, with the very body which was marked by the nails upon His cross, and wherewith He ate and drank after His resurrection—that He who repelled the imagination of His disciples, as if they had seen a spirit, by bidding them handle Him and see, and subjecting to their familiar touch, the flesh and the bones that encompassed Him; that He should now be throned in universal supremacy, and wielding the whole power of heaven and earth, have every knee to bow at His name, and every tongue to confess, and yet all to the glory of God the Father—that humanity, that substantial and embodied humanity, should thus be exalted, and a voice of adoration from every creature, be lifted up to the Lamb for ever and ever—does this look like the abolition of materialism, after the present system of it is destroyed; or does it not rather prove, that transplanted into another system, it will be preferred to celestial honours, and prolonged in immortality throughout all ages?

It has been our careful endeavour, in all that we have said, to keep within the limits of the record, and to offer no other remarks than those which may fitly be suggested by the circumstance, that a new earth is to be created, as well as a new heavens, for the future accommodation of the righteous. We have no desire to push the speculation beyond what is written—but it were, at the same time, well, that in all our representations of the immortal state, there was just the same force of colouring, and the same vivacity of scenic exhibition, that there is in the New Testament. The imagination of a total and diametric opposition between the region of sense and the region of spirituality, certainly tends to abate the interest with which we might otherwise look to the perspective that is on the other side of the grave; and to deaden all those sympathies that we else might have with the joys and the exercises of the blest in paradise. To rectify this, it is not necessary to enter on the particularities of heaven—a topic on which the Bible is certainly most sparing and reserved in its communications. But a great step is gained, simply by dissolving the alliance that exists in the minds of many between the two ideas of sin and materialism; or proving, that when once sin is done away, it consists with all we know of God's administration, that materialism shall be perpetuated in the full bloom and vigour of immortality. It altogether holds out a warmer and more alluring picture of the elysium that awaits us, when told, that there, will be beauty to delight the eye; and music to regale the ear; and the comfort that springs from all the charities of intercourse between man and man, holding converse as they do on earth, and gladdening each other with the benignant smiles that play on the human countenance, or the accents of kindness that fall in soft and soothing melody from the human voice. There is much of the innocent, and much of the inspiring, and much to affect and elevate the heart, in the scenes and the contemplations of materialism—and we do hail the information of our text, that after the dissolution of its present frame-work, it will again be varied and decked out anew in all the graces of its unfading verdure, and of its unbounded variety—that in addition to our direct and personal view of the Deity, when He comes down to tabernacle with men, we shall also have the reflection of Him in a lovely mirror of His own workmanship—and that instead of being transported to some abode of dimness and of mystery, so remote from human experience, as to be beyond all comprehension, we shall walk for ever in a land replenished with those sensible delights, and those sensible glories, which, we doubt not, will lie most profusely scattered over the "new heavens and the new earth, wherein dwelleth righteousness."

II. But though a paradise of sense, it will not be a paradise of sensuality. Though not so unlike the present world as many apprehend it, there will be one point of total dissimilarity betwixt them. It is not the entire substitution of spirit for matter, that will distinguish the future economy from the present. But it will be the entire substitution of righteousness for sin. It is this which signalizes the Christian from the Mahometan paradise—not that sense, and substance, and splendid imagery, and the glories of a visible creation seen with bodily eyes, are excluded from it,—but that all which is vile in principle, or voluptuous in impurity, will be utterly excluded from it. There will be a firm earth, as we have at present, and a heaven stretched over it, as we have at present; and it is not by the absence of these, but by the absence of sin, that the abodes of immortality will be characterized. There will both be heavens and earth, it would appear, in the next great administration—and with this specialty to mark it from the present one, that it will be a heavens and an earth, "wherein dwelleth righteousness."

Now, though the first topic of information that we educed from the text, may be regarded as not very practical, yet the second topic on which we now insist, is most eminently so. Were it the great characteristic of that spirituality which is to obtain in a future heaven, that it was a spirituality of essence, then occupying and pervading the place from which materialism had been swept away, we could not, by any possible method, approximate the condition we are in at present, to the condition we are to hold everlastingly. We cannot etherealize the matter that is around us—neither can we attenuate our own bodies, nor bring down the slightest degree of such a heaven to the earth that we now inhabit. But when we are told that materialism is to be kept up, and that the spirituality of our future state lies not in the kind of substance which is to compose its framework, but in the character of those who people it—this puts, if not the fulness of heaven, at least a foretaste of heaven, within our reach. We have not to strain at a thing so impracticable, as that of diluting the material economy which is without us—we have only to reform the moral economy that is within us. We are now walking on a terrestrial surface, not more compact, perhaps, than the one we shall hereafter walk upon, and are now wearing terrestrial bodies, not firmer and more solid, perhaps, than those we shall hereafter wear. It is not by working any change upon them, that we could realize, to any extent, our future heaven. And this is simply done by opening the door of our heart for the influx of heaven's affections—by bringing the whole man, as made up of soul, and spirit, and body, under the presiding authority of heaven's principles.

This will make plain to you how it is, that it could be said in the New Testament, that the "kingdom of heaven was at hand"—and how, in that book, its place is marked out, not by locally pointing to any quarter, and saying, Lo, here, or lo there, but by the simple affirmation that the kingdom of heaven is within you—and how, in defining what it was that constituted the kingdom of heaven, there is an enumeration, not of such circumstances as make up an outward condition, but of such feelings and qualities as make up a character, even righteousness, and peace, and joy in the Holy Ghost—and how the ushering in of the new dispensation is held equivalent to the introduction of this kingdom into the world—all making it evident, that if the purity and the principles of heaven begin to take effect upon our heart, what is essentially heaven begins with us, even in this world; that instead of ascending to some upper region, for the purpose of entering it, it may descend upon us, and make an actual entrance of itself into our bosoms; and that so far, therefore, from that remote and inaccessible thing which many do regard it, it may, through the influence of the word which is nigh unto you, and of the Spirit that is given to prayer, be lighted up in the inner man of an individual upon earth, whose person may even here, exemplify ts graces, and whose soul may even here realize a measure of its enjoyments.

And hence one great purpose of the incarnation of our Saviour. He came down amongst us in the full perfection of heaven's character, and has made us see, that it is a character which may be embodied. All its virtues were, in his case, infused into a corporeal frame-work, and the substance of these lower regions was taken into intimate and abiding association with the spirit of the higher. The ingredient which is heavenly, admits of being united with the ingredient which is earthly—so that we, who, by nature, are of the earth, and earthly, could we catch of that pure and celestial element which made the man Christ Jesus to differ from all other men, then might we too be formed into that character, by which it is that the members of the family above differ from those of the outcast family beneath. Now, it is expressly said of Him, that He is set before us as an example; and we are required to look to that living exhibition of Him, where all the graces of the upper sanctuary are beheld as in a picture; and instead of an abstract, we have in His history a familiar representation of such worth, and piety, and excellence, as could they only be stamped upon our own persons, and borne along with us to the place where He now dwelleth—instead of being shunned as aliens, we should be welcomed and recognized as seemly companions for the inmates of that place of holiness. And, in truth, the great work of Christ's disciples upon earth, is a constant and busy process of assimilation to their Master who is in heaven. And we live under a special economy, that has been set up for the express purpose of helping it forward. It is for this, in particular, that the Spirit is provided. We are changed into the image of the Lord, even by the Spirit of the Lord. Nursed out of this fulness, we grow up unto the stature of perfect men in Christ Jesus—and instead of heaven being a remote and mysterious unknown, heaven is brought near to us by the simple expedient of inspiring us where we now stand, with its love, and its purity, and its sacredness. We learn from Christ, that the heavenly graces are all of them compatible with the wear of an earthly body, and the circumstances of an earthly habitation. It is not said in how many of its features the new earth will differ from, or be like unto the present one—but we, by turning from our iniquities unto Christ, push forward the resemblance of the one to the other, in the only feature that is specified, even that "therein dwelleth righteousness."

And had we only the character of heaven, we should not be long of feeling what that is which essentially makes the comfort of heaven. "Thou lovest righteousness, and hatest iniquity; therefore, God thy God, hath anointed thee with the oil of gladness, above thy fellows." Let us but love the righteousness which He loves, and hate the iniquity which He hateth; and this, of itself, would so soften and attune the mechanism of our moral nature, that in all the movements of it, there should be joy. It is not sufficiently adverted to, that the happiness of heaven lies simply and essentially in the well-going machinery of a well-conditioned soul—and that according to its measure, it is the same in kind with the happiness of God, who liveth for ever in bliss ineffable, because He is unchangeable in being good, and upright, and holy. There may be audible music in heaven, but its chief delight will be in the music of well-poised affections, and of principles in full and consenting harmony with the laws of eternal rectitude. There may be visions of loveliness there; but it will be the loveliness of virtue, as seen directly in God, and as reflected back again in family likeness from all His children—it will be this that shall give its purest and sweetest transports to the soul. In a word, the main reward of paradise, is spiritual joy—and that, springing at once from the love and the possession of spiritual excellence. It is such a joy as sin extinguishes on the moment of its entering the soul, and such a joy as is again restored to the soul, and that immediately on its being restored to righteousness.

It is thus that heaven may be established upon earth, and the petition of our Lord's prayer be fulfilled, "Thy kingdom come." This petition receives its best explanation from the one which follows: "Thy will be done on earth as it is done in heaven." It just requires a similarity of habit and character in the two places, to make out a similarity of enjoyment. Let us attend, then, to the way in which the services of the upper sanctuary are rendered—not in the spirit of legality, for this gendereth to bondage; but in the spirit of love, which gendereth to the beatitude of the affections rejoicing in their best and most favourite indulgence. They do not work there, for the purpose of making out the conditions of a bargain. They do not act agreeably to the pleasure of God, in order to obtain the gratification of any distinct will or distinct pleasure of their own, in return for it. Their will is, in fact, identical with the will of God. There is a perfect unison of taste and of inclination, between the creature and the Creator. They are in their element, when they are feeling righteously, and doing righteously. Obedience is not drudgery, but delight to them; and as much as there is of the congenial between animal nature, and the food that is suitable to it, so much is there of the congenial between the moral nature of heaven, and its sacred employments and services. Let the will of God, then, be done here, as it is done there, and not only will character and conduct be the same here as there, but they will also resemble each other in the style, though not in the degree of their blessedness. The happiness of heaven will be exemplified upon earth, along with the virtue of heaven—for, in truth, the main ingredient of that happiness is not given them in payment for work; but it lies in the love they bear to the work itself. A man is never happier than when employed in that which he likes best. This is all a question of taste: but should such a taste be given as to make it a man's meat and drink to do the will of his Father, then is he in perfect readiness for being carried upwards to heaven, and placed beside the pure river of water of life, that proceedeth out of the throne of God and of the Lamb. This is the way in which you may make a heaven upon earth, not by heaping your reluctant offers at the shrine of legality, but by serving God because you love him; and doing his will, because you delight to do him honour.

And here we may remark, that the only possible conveyance for this new principle into the heart, is the Gospel of Jesus Christ,—that in no other way than through the acceptance of its free pardon, sealed by the blood of an atonement, which exalts the Lawgiver, can the soul of man be both emancipated from the fear of terror, and solemnized into the fear of humble and holy reverence—that it is only in conjunction with the faith that justifies, that the love of gratitude, and the love of moral esteem, are made to arise in the bosom of regenerated man; and, therefore, to bring down the virtues of heaven, as well as the peace of heaven, into this lower world, we know not what else can be done, than to urge upon you the great propitiation of the New Testament—nor are we aware of any expedient by which all the cold and freezing sensations of legality can be done away, but by your thankful and unconditional acceptance of Jesus Christ, and him crucified.

DISCOURSE V.

THE NATURE OF THE KINGDOM OF GOD.

"For the kingdom of God is not in word, but in power."—
1 CORINTHIANS iv. 20.

THERE is a most important lesson to be derived from the variety of senses in which the phrases "kingdom of God," and "kingdom of heaven," are evidently made use of in the New Testament. If it, at one time, carry our thoughts to that place where God sits in visible glory, and where, surrounded by the family of the blessed, he presides in full and spiritual authority—it, at another time, turns our thoughts inwardly upon ourselves, and instead of leading us to say, Lo, here, or lo, there, as if to some local habitation at a distance, it leads us, by the declaration, that the "kingdom of God is within us," to look for it into our own breast, and to examine whether heavenly affections have been substituted there in the place of earthly ones. Such is the tendency of our imagination upon this subject, that the kingdom of heaven is never mentioned, without our minds being impelled thereby to take an upward direction—to go aloft to that place of spaciousness, and of splendour, and of psalmody, which forms the residence of angels; and where the praises both of redeemed and unfallen creatures, rise in one anthem of gratulation to the

Father, who rejoices over them all. Now, it is evident, that in dwelling upon such an elysium as this, the mind can picture to itself a thousand delicious accompaniments, which, apart from moral and spiritual character altogether, are fitted to regale animal, and sensitive, and unrenewed man. There may be sights of beauty and brilliancy for the eye. There may be sounds of sweetest melody for the ear. There may be innumerable sensations of delight, from the adaptation which obtains between the materialism of surrounding heaven, and the materialism of our own transformed and glorified bodies. There may even be poured upon us, in richest abundance, a higher and a nobler class of enjoyments —and separate still from the possession of holiness, of that peculiar quality, by the accession of which a sinner is turned into a saint, and the man who, before, had an entire aspect of secularity and of the world, looks as if he had been cast over again in another mould, and come out breathing godly desires, and aspiring, with a newly created fervour, after godly enjoyments. And so, without any such conversion as this, heaven may still be conceived to minister a set of very refined and intellectual gratifications. One may figure it so formed, as to adapt itself to the senses of man, though he should possess not one single virtue of the temple, or of the sanctuary—and one may figure it to be so formed, as, though alike destitute of these virtues, to adapt itself even to the Spirit of man, and to many of the loftier principles and capacities of his nature. His taste may find an ever-recurring delight in the panorama of its sensible glories; and his fancy wander untired among all the realities and all the possibilities of created excellence; and his understanding be feasted to ecstasy among those endless varieties of truth, which are ever pouring in a rich flood of discovery, upon his mind; and even his heart be kept in a glow of warm and kindly affection among the cordialities of that benevolence, by which he is surrounded. All this is possible to be conceived of heaven—and when we add its secure and everlasting exemption from the agonies of hell, let us not wonder, that such a heaven should be vehemently desired by those who have not advanced by the very humblest degree of spiritual preparation, for the real heaven of the New Testament—who have not the least congeniality of feeling with that which forms its essential and characteristic blessedness—who cannot sustain on earth for a very short interval of retirement, the labour and the weariness of communion with God—who, though they could relish to the uttermost, all the sensible and all the intellectual joys of heaven, yet hold no taste of sympathy whatever, with its hallelujahs, and its songs of raptured adoration—and who, therefore, if transported at this moment, or if transported after death, with the frame and character of soul that they have at this moment, to the New Jerusalem, and the city of the living God, would positively find themselves aliens, and out of their kindred and rejoicing element, however much they may sigh after a paradise of pleasure, or a paradise of poetry.

It may go to dissipate this sentimental illusion, if we ponder well the meaning which is often assigned to the kingdom of heaven in the Bible—if we reflect, that it is often made to attach personally to a human creature upon earth—as well as to be situated locally in some distant and mysterious region away from us—that to be the subjects of such a kingdom, it is not indispensable that our residence be within the limits of an assigned territory, any more, in fact, than that the subject of an earthly sovereign should not remain so, though travelling, for a time, beyond the confines of his master's jurisdiction. He may, though away from his country in person, carry about with him in mind a full principle of allegiance to his country's sovereign—and may both, in respect of legal duty, and of his own most willing and affectionate compliance with it, remain associated with him both in heart and in political relationship. He is still a member of that kingdom, in the domains of which he was born—and in the very same way, may a man be travelling the journey of life in this world, and be all the while a member of the kingdom of heaven. The Being who reigns in supreme authority there, may, even in this land of exile and alienation, have some one devoted subject, who renders to the same authority the deference of his heart, and the subordination of his whole practice. The will of God may possess such a moral ascendancy over his will, as that when the one commands, the other promptly and cheerfully obeys. The character of God may stand revealed in such charms of perfection and gracefulness to the eye of his mind, that by ever looking to Him, he both loves and is made like unto Him. A sense of God may pervade his every hour, and every employment, even as it is the hand of God which preserves him continually, and through the actual power of God, that he lives and moves, as well as has his being. Such a man, if such a man there be on the face of our world, has the kingdom of God set up in his heart. He is already one of the children of the kingdom. He is not locally in heaven, and yet his heaven is begun. He has in his eye the glories of heaven; though, as yet, he sees them through a glass darkly. He feels in his bosom the principles of heaven; though still at war with the propensities of nature, they do not yet reign in all the freeness of an undisputed ascendancy. He carries in his heart the peace, and the joy, and the love, and the elevation of heaven; though, under the incumbrance of a vile body, the spiritual repast which is thus provided, is not without its mixtures, and without its mitigation. In a word, the essential elements of heaven's reward, and of heaven's felicity, are all in his possession. He tastes the happiness of heaven in kind, though not in its full and finished degree. When he gets to heaven above, he will not meet there with a happiness differing in character from that which he now feels; but only higher in gradation. There may be crowns of material splendour. There may be trees of unfading loveliness. There may be pavements of emerald—and canopies of brightest radiance—and gardens of deep and tranquil security—and palaces of proud and stately decoration—and a city of lofty pinnacles, through which there unceasing flows a river of gladness, and where jubilee is ever rung with the concord of seraphic voices. But these are only the accessaries of heaven. They form not the materials of its substantial blessedness. Of this the man who toils in humble drudgery, an utter stranger to the delights of sensible pleasure, or the fascinations of sensible glory, has got already a foretaste in his heart. It consists not in the enjoyment of created good, nor in the survey of created magnificence. It is drawn in a direct stream, through the channels of love and of contemplation, from the fulness of the Creator. It emanates from the countenance of God, manifesting the spiritual glories of His holy and perfect character, on those whose characters are kindred to His own. And if on earth there is no tendency towards such a character—no process of restoration to the lost image of the Godhead—no delight in prayer—no relish for the sweets of intercourse with our Father, now unseen, but then to be revealed to the view of His immediate worshippers—then, let our imaginations kindle as they may, with the beatitudes of our fictitious heaven, the true heaven of the Bible is what we shall never reach, because it is a heaven that we are not fitted to enjoy.

But such a view of the matter seems not merely to dissipate a sentimental illusion which obtains upon this subject. It also serves to dissipate a theological illusion. Ere we can enter heaven, there must be granted to us a legal capacity of admission—and Christ by His atoning death, and perfect righteousness, has purchased this capacity for those who believe—and they, by the very act of believing, are held to be in possession of it, just as a man by stretching out his hand to a deed or a passport, becomes vested with all the privileges which are thereby conveyed to the holder. Now, in the zeal of controversialists, (and it is a point most assuredly about which they cannot be too zealous)—in their zeal to clear up and to demonstrate the ground on which the sinner's legal capacity must rest—there has, with many, been a sad overlooking of what is no less indispensable, even his personal capacity. And yet even on the lowest and grossest conceptions of what that is which constitutes the felicity of heaven, it would be no heaven, and no place of enjoyment at all, without a personal adaptation on the part of its occupiers, to the kind of happiness which is current there. If that happiness consisted entirely in sights of magnificence, of what use would it be to confer a title-deed of entry on a man who was blind? To make it heaven to him, his eyes must be opened. Or, if that happiness consisted in sounds of melody, of what use would a passport be to the man who was deaf? To make out a heaven for him, a change must be made on the person which he wears, as well as in the place which he occupies—and his ears must be unstopped. Or, if that happiness consisted in fresh and perpetual accessions of new and delightful truth to the understanding, what would rights and legal privileges avail to him who was sunk in helpless idiotism? To provide him with a heaven, it is not enough that he be transported to a place among the mansions of the celestial: he must be provided with a new faculty—and, as before, a change behoved to be made upon the senses; so now, ere heaven can be heaven to its occupier, a change must be made upon his mind. And, in like manner, my brethren, if that happiness shall consist in the love of God for His goodness, and in the love of God for the moral and spiritual excellence which belongs to Him—if it shall consist in the play and exercise of affections directed to such objects as are alone worthy of their most exalted regard—if it shall consist in the movements of a heart now attracted in reverence and admiration towards all that is noble, and righteous, and holy—it is not enough to constitute a heaven for the sinner, that God is there in visible manifestation, or that heaven is lighted up to him in a blaze of spiritual glory. His heart must be made a fit recipient for the impression of that glory. Of what possible enjoyment to him is heaven, as his purchased inheritance, if heaven be not also his precious and his much-loved home? To create enjoyment for a man, there must be a suitableness between the taste that is in him, and the objects that are around him. To make a natural man happy upon earth, we may let his taste alone, and surround him with favourable circumstances—with smiling abundance, and merry companionship, and bright anticipations of fortune or of fame, and the salutations of public respect, and the gaieties of fashionable amusement, and the countless other pleasures of a world, which yields so much to delight and to diversify the short-lived period of its fleeting generations. To make the same man happy in heaven, it would suffice simply to transmit him there with the same taste, and to surround him with the same circumstances. But God has not so ordered heaven. He will not suit the circumstance of heaven to the character of man—and therefore to make it, that man can be happy there, nothing remains but to suit the character of man to the circumstances of heaven—and, therefore it is, that to bring about heaven to a sinner, it is not enough that there be the preparation of a place for him, there must be a preparation of him for the place—it is not enough that he be meet in law, he must be meet in person—it is not enough that there be a change in his forensic relation towards God, there must be a change in the actual disposition of his heart towards Him; and unless delivered from his earth-born propensities—unless a clean heart be created, and a right spirit be renewed—unless transformed into a holy and a godlike character, it is quite in vain to have put a deed of entry into his hands—heaven will have no charm for him—all its notes of rapture will fall with tasteless insipidity upon his ear—and justification itself will cease to be a privilege.

Let us cease to wonder, then, at the frequent application, in Scripture, of this phrase to a state of personal feeling and character upon earth—and rather let us press upon our remembrance the important lessons which are to be gathered from such an application. In that passage where it is said, that the "kingdom of God is not meat and drink, but righteousness, and peace, and joy in the Holy Ghost," there can be no doubt that the reference is altogether personal, for the apostle is here contrasting the man who, in these things, serveth Christ, with the man who eateth unto the Lord, or who eateth not unto the Lord. And in the passage now before us, there can be as little doubt, that the reference is to the kingdom of God, as fixed and substantiated upon the character of the human soul. He was just before alluding to those who could talk of the things of Christ, while it remained questionable whether there was any change or any effect that could at all attest the power of these things upon their person and character. This is the point which he proposed to ascertain on his next visit to them. "I will come to you shortly, if the Lord will, and will know not the speech of them which are puffed up, but the power. For the kingdom of God is not in word, but in power." It is not enough to mark you as the children of this kingdom; or as those over whose hearts the reign of God is established; or as those in whom a preparation is going on here for a place of glory and blessedness hereafter—that you know the terms of orthodoxy, or that you can speak its language. If even an actual belief in its doctrine could reside in your mind, without fruit and without influence, this would as little avail you. But it is well to know, both from experience and from the information of Him who knew what was in man, that an actual belief of the Gospel, is at all times an effectual belief—that upon the entrance of such a belief, the kingdom of God comes to us with power, being that which availeth, even faith working by love, and purifying the heart, and overcoming the world.

One of the simplest cases of the kingdom of God in word, and not in power, is that of a child, with its memory stored in passages of Scripture, and in all the answers to all the questions of a substantial and well-digested catechism. In such an instance, the tongue may be able to rehearse the whole expression of evangelical truth, while neither the meaning of the truth is perceived by the understanding, nor of consequence, can the moral influence of the truth be felt in the heart. The learner has got words, but nothing more. This is the whole fruit of his acquisition—nor would it make any difference, in as far as the effect at the time is concerned, though, instead of words adapted to the expression of Christian doctrine, they had been the words of a song, or a fable, or any secular narrative and performance whatever. This is all undeniable enough—if we could only prevail on many men, and many women, not to deny its application to themselves—if we could only convince our grown-up children of the absolute futility of many of their exercises—if we could only arouse from their dormancy, our listless readers of the Bible—our men, who make a mere piece-work of their Christianity; who, in making way through the Scriptures, do it by the page, and, in addressing prayers to their Maker, do it by the sentence; with whom the perusal of the sacred volume, is absolutely little better than a mere exercise of the lip, or of the eye, and a preference for orthodoxy is little better than a preference for certain familiar and well-known sounds; where the thinking principle is almost never in contact with the matter of theological truth, however conversant both their mouths and their memories may be with the language of it—so that in fact the doctrine by the knowledge of which, and the power of which it is, that we are saved, lies as effectually hidden from their minds, as if it lay wrapt in hieroglyphical obscurity; or, as if their intellectual organ was shut against all communication with any thing without them—and thus it is, that what is not perceived by the mental eye, having no possible operation upon the mental feelings, or mental purposes, the kingdom of God cometh to them in word only, while not in power.

But again, what is translated word in this verse, is also capable of being rendered by the term reason. It may not only denote that which constitutes the material vehicle by which the argument conceived in the mind of one man is translated into the mind of another—it may also denote the argument itself; and when rendered in this way, it offers to our notice a very interesting case, of which there are not wanting many exemplifications. In the case just now adverted to, the mere word is in the mouth, without its corresponding idea being in the mind; but in the case immediately before us, ideas are present as well as words, and every intellectual faculty is at its post, for the purpose of entertaining them—the attention most thoroughly awake—and the curiosity on the stretch of its utmost eagerness—and the judgment most busily employed in the work of comparing one doctrine, and one declaration with another—and the reason conducting its long or its intricate processes—and, in a word, the whole machinery of the mind as powerfully stimulated by a theological, as it ever can be, by a natural or scientific speculation—and yet, with this seeming advancement that it makes from the language of Christianity to the substance of Christianity, what shall we think of it, if there be no advancement whatever in the power of Christianity—no accession to the soul of any one of those three ingredients, which, taken together, make up the apostle's definition of the kingdom of God—no augmentation either of its righteousness, or its peace, or its joy in the Holy Ghost—the man, no doubt, very much engrossed and exercised with the subject of divinity, but with as little of the real spirit and character of divinity, thereby transferred into his own spirit, and his own character, as if he were equally engrossed and equally exercised with the subject of mathematics—remaining in short, after all his doctrinal acquisitions of the truth, an utter stranger to the moral influence of the truth—and proving, in the fact of his being practically and personally the very same man as before, that if the kingdom of God is not in word, it is as little in argument, but in power.

If it be of importance to know, that a man may lay hold, by his memory, of all the language of Christianity, and yet not be a Christian—it is also of importance to know, that a man may lay hold, by his understanding, of all the doctrine of Christianity, and yet not be a Christian. It is our opinion, that in this case the man has only an apparent belief, without having an actual belief—that all the doctrine is conceived by him, without being credited by him—that it is the object of his fancy, without being the object of his faith—and that, as on the one hand, if the conviction be real, the consequence of another heart, and another character, will be sure—so, on the other hand, and on the principle of "by their fruits shall ye know them," if he want the fruit, it is just because he is in want of the foundation--if there be no produce, it is because there is no principle—having experienced no salvation from sin here, he shall experience no salvation from the abode of sinners hereafter. If faith were present with him, he would be kept by the power of it unto salvation, from both—but destitute as he proves himself to be now of the faith which sanctifies, he will be found then, in the midst of all his semblances and all his delusions, to have been equally destitute of the faith which justifies.

And it is, perhaps, not so difficult to stir up, in the mind of the learned controversialist, and the deeply-exercised scholar, the suspicion, that with all his acquirements in the lore of theology, he is, in respect of its personal influence upon himself, still in a state of moral and spiritual unsoundness—it is not so difficult to raise this feeling of self-condemnation in his mind, as it is to do it in the mind of him who has selected his one favourite article, and there, resolved, if die he must, to die hard, has taken up his obstinate and immoveable position—and retiring within the intrenchment of a few verses of the Bible, will defy all the truth and all the thunder of its remaining declarations; and with an orthodoxy which carries on all its play in his head, without one moving or one softening touch upon his heart, will stand out to the eye of the world, both in avowed principle, and in its corresponding practice, a secure, sturdy, firm, impregnable Antinomian. He thinks that he will have heaven, because he has faith. But if his faith do not bring the virtues of heaven into his heart, it will never spread either the glory or the security of heaven around his person. The region to which he vainly thinks of looking forward, is a region of spirituality—and he himself must be spiritualized, ere it can prove to him a region of enjoyment. If he count on a different paradise from this, he is as widely mistaken as they who dream of the luxury that awaits them in the paradise of Mahomet. He misinterprets the whole undertaking of Jesus Christ. He degrades the salvation which he hath achieved, into a salvation from animal pain. He transforms the heaven which He has opened, into a heaven of animal gratifications. He forgets, that on the great errand of man's restoration, it is not more necessary to recall our departed species to the heaven from which they had wandered, than it is to recall to the bosom of man its departed worth, and its departed excellence. The one is what faith will do on the other side of death. But the other just as certainly faith must do on this side of death. It is here that heaven begins. It is here that eternal life is entered upon. It is here that man first breathes the air of immortality. It is upon earth that he learns the rudiments of a celestial character, and first tastes of celestial enjoyments. It is here, that the well of water is struck out in the heart of renovated man, and that fruit is made to grow unto holiness, and then, in the end, there is life everlasting. The man whose threadbare orthodoxy is made up of meagre and unfruitful positions, may think that he walks in clearness, while he is only walking in the cold light of speculation He walks in the feeble sparks of his own kindling. Were it fire from the sanctuary, it would impart to his unregenerated bosom, of the heat, and spirit, and love of the sanctuary. This is the sure result of the faith that is unfeigned—and all that a feigned faith can possibly make out, will be a fictitious title-deed, which will not stand before the light of the great day of final examination. And thus will it be found, I fear, in many cases of marked and ostentatious professorship, how possible a thing it is to have an appearance of the kingdom of God in word, and the kingdom of God in letter, and the kingdom of God in controversy—while the kingdom of God is not in power.

But once more—instead of laying a false security upon one article, it is possible to have a mind familiarized to all the articles—to admit the need of holiness, and to demonstrate the channel of influence by which it is brought down from heaven upon the hearts of believers—to cast an eye of intelligence over the whole symphony and extent of Christian doctrine—to lay bare those ligaments of connexion by which a true faith in the mind is ever sure to bring a new spirit and a new practice along with it—and to hold up the lights both of Scripture and of experience, over the whole process of man's regeneration. It is possible for one to do all this—and yet to have no part in that regeneration—to declare with ability and effect the Gospel to others, and yet himself be a castaway—to unravel the whole of that spiritual mechanism, by which a sinner is transformed into a saint, while he does not exemplify the working of that mechanism in his own person—to explain what must be done, and what must be undergone in the process of becoming one of the children of the kingdom, while he himself remains one of the children of this world. To him the kingdom of God hath come in word, and it hath come in letter, and it hath come in natural discernment; but it hath not come in power. He may have profoundly studied the whole doctrine of the kingdom—and have conceived the various ideas of which it is composed—and have embodied them in words—and have poured them forth in utterance—and yet be as little spiritualized by these manifold operations, as the air is spiritualized by its being the avenue for the sounds of his voice to the ears of his listening auditory. The living man may, with all the force of his active intelligence, be a mere vehicle of transmission. The Holy Ghost may leave the message to take its own way through his mind—and may refuse the accession of His influence, till it make its escape from the lips of the preacher—and may trust for its conveyance to those aërial undulations by which the report is carried forward to an assembled multitude—and may only, after the entrance of hearing has been effected for the terms of the message, may only, after the unaided powers of moral and physical nature have brought the matter thus far, may then, and not till then, add His own influence to the truths of the message, and send them with this impregnation from the ear to the conscience of any whom He listeth. And thus from the workings of a cold and desolate bosom in the human expounder, may there pro ceed a voice, which on its way to some of those who are assembled around him, shall turn out to be a voice of urgency and power. He may be the instrument of blessings to others, which have never come with kindly or effective influence upon his own heart. He may inspire an energy, which he does not feel, and pour a comfort into the wounded spirit, the taste of which, and the enjoyment of which is not permitted to his own—and nothing can serve more effectually than this experimental fact to humble him, and to demonstrate the existence of a power which cannot be wielded by all the energies of Nature—a power often refused to eloquence, often refused to the might and the glory of human wisdom—often refused to the most strenuous exertions of human might and human talent, and generally met with in richest abundance among the ministrations of the men of simplicity and prayer.

Some of you have heard of the individual who, under an oppression of the severest melancholy, implored relief and counsel from his physician. The unhappy patient was advised to attend the performances of a comedian, who had put all the world into ecstasies. But it turned out, that the patient was the comedian himself—and that while his smile was the signal of merriment to all, his heart stood uncheered and motionless, amid the gratulations of an applauding theatre—and evening after evening, did he kindle around him a rapture in which he could not participate—a poor, helpless, dejected mourner, among the tumults of that high-sounding gaiety, which he himself had created.

Let all this touch our breasts with the persuasion, of the nothingness of man. Let it lead us to withdraw our confidence from the mere instrument, and to carry it upwards to Him who alone worketh all in all. Let it reconcile us to the arrangements of His providence, and assure our minds, that He can do with one arrangement, what we fondly anticipated from another. Let us cease to be violently affected by the mutabilities of a fleeting and a shifting world—and let nothing be suffered to have the power of dissolving for an instant, that connection of trust which should ever subsist between our minds and the will of the all-working Deity. Above all, let us carefully separate between our liking for certain accompaniments of the word, and our liking for the word itself. Let us be jealous of those human preferences, which may bespeak some human and adventitious influence upon our hearts, and be altogether different from the influence of Christian truth upon Christianized and sanctified affections. Let us be tenacious only of one thing—not of holding by particular ministers—not of saying, that "I am of Paul, or Cephas, or Apollos"—not of idolizing the servant, while the Master is forgotten,—but let us hold by the Head, even Christ. He is the source of all spiritual influence—and while the agents whom he employs, can do no more than bring the kingdom of God to you in word—it lies with him either to exalt one agency, or to humble and depress another

—and either with or without such an agency, by the demonstration of that Spirit, which is given unto faith, to make the kingdom of God come into your hearts with power.

DISCOURSE VI.

HEAVEN A CHARACTER AND NOT A LOCALITY.

"He that is unjust, let him be unjust still: and he which is filthy, let him be filthy still: and he that is righteous, let him be righteous still: and he that is holy, let him be holy still." —Rev. xxii. 11.

Our first remark on this passage of Scripture, is, how very palpably and nearly it connects time with eternity. The character wherewith we sink into the grave at death, is the very character wherewith we shall re-appear on the day of resurrection. The character which habit has fixed and strengthened through life, adheres, it would seem, to the disembodied spirit, through the mysterious interval which separates the day of our dissolution from the day of our account—when it will again stand forth, the very image and substance of what it was, to the inspection of the Judge and the awards of the judgment-seat. The moral lineaments which be graven on the tablet of the inner man, and which every day of an unconverted life makes deeper and more indelible than before, will retain the very impress they have gotten—unaltered and uneffaced, by the transition from our present to our future state of existence. There will be a dissolution, and then a reconstruction of the body, from the sepulchral dust into which it had mouldered. But there will be neither a dissolution nor a renovation of the spirit, which, indestructible both in character and essence, will weather and retain its identity, on the mid-way passage between this world and the next—so that at the time of quitting its earthly tenement we may say, that, if unjust now it will be unjust still, if filthy now it will be filthy still, if righteous now it will be righteous still, and if holy now it will be holy still.

Our second remark, suggested by the scripture now under consideration, is that there be many analogies of nature and experience, which even death itself does not interrupt. There is nought more familiar to our daily observation than the power and inveteracy of habit—insomuch that any vicious propensity is strengthened by every new act of indulgence; any virtuous principle is more firmly established than before, by every new act of resolute obedience to its dictates. The law which connects the actings of boyhood or of youth with the character of manhood, is the identical, the unrepealed law which connects our actings in time with our character through eternity. The way in which the moral discipline of youth prepares for the honours and the enjoyments of a virtuous manhood, is the very way in which the moral and spiritual discipline of a whole life prepares for a virtuous and happy immortality. And, on the other hand, the succession, as of cause and effect, from a profligate youth or a dishonest manhood, to a disgraced and worthless old age—is just the succession, also of cause and effect, between the misdeeds and the depravities of our history on earth, and an inheritance of worthlessness and wretchedness for ever. The law of moral continuity between the different stages of human life, is also the law of continuity between the two worlds—which even the death that intervenes does not violate. Be he a saint or a sinner, each shall be filled with the fruit of his own ways—so that when translated into their respective places of fixed and everlasting destination, the one shall rejoice through eternity in that pure element of goodness, which here he loved and aspired after; the other, a helpless, a degraded victim of those passions which lorded over him through life, shall be irrevocably doomed to that worst of torments and that worst of tyranny —the torment of his own accursed nature, the inexorable tyranny of evil.

Our third remark suggested by this scripture is, that it affords no very dubious perspective of the future heaven and the future hell of the New Testament. We are aware of the material images employed in scripture, and by which it bodies forth its representation of both—of the fire, and the brimstone, and the lake of living agony, and the gnashing of teeth, and the wailings, the ceaseless wailings of distress and despair unutterable, by which the one is set before us in characters of terror and most revolting hideousness—of the splendour, the spaciousness, the music, the floods of melody and sights of surpassing loveliness, by which the other is set before us in characters of bliss and brightness unperishable; with all that can regale the glorified senses of creatures, rejoicing for ever in the presence and before the throne of God. We stop not to inquire, and far less to dispute, whether these descriptions, in the plain meaning and very letter of them, are to be realized. But we hold that it would purge theology from many of its errors, and that it would guide and enlighten the practical Christianity of many honest inquirers—if the moral character both of heaven and hell were more distinctly recognized, and held a more prominent place in the regards and contemplations of men. If it indeed be true that the moral, rather than the material, is the main ingredient, whether of the coming torment or the coming ecstasy—then the hell of the wicked may be said to have already begun, and the heaven of the virtuous may be said to have already begun. The one, in the bitterness of an unhinged and dissatisfied spirit, has a foretaste of the wretchedness before him; the other, in the peace and triumphant complacency of an approving conscience, has a foretaste of the happiness before him. Each is ripening for his own everlasting destiny; and whether in the depravities that deepen and accumulate on the character of the one, or in the graces that brighten and multiply upon the other—we see materials enough, either for the worm that dieth not, or for the pleasures that are for evermore.

But again, it may be asked, will spiritual elements alone suffice to make up, either the intense and intolerable wretchedness of a hell, or the intense beatitude of a heaven? For an answer to this question, let us first turn your attention to the former of these receptacles. And we ask you to think of the state of that heart in respect to sensation, which is the seat of a concentrated and all-absorbing selfishness, which feels for no other interest than its own, and holds no fellowship of truth or honesty or confidence with the fellow-beings around it. The owner of such a heart may live in society; but, cut off as he is by his own sordid nature from the reciprocities of honourable feeling and good faith, he may be said to live an exile in the midst of it. He is a stranger to the day-light of the moral world; and, instead of walking abroad on an open platform of free and fearless communion with his fellows, he spends a cold and heartless existence in the hiding-place of his own thoughts. You mistake it, if you think of this creeping and ignoble creature, that he knows aught of the real truth or substance of enjoyment; or however successful he may have been in the wiles of his paltry selfishness, that a sincere or a solid satisfaction has been the result of it. On the contrary, if you enter his heart, you will there find a distaste and disquietude in the lurking sense of its own worthlessness; and that dissevered from the respect of society without, it finds no refuge within where he is abandoned by the respect of his own conscience. It does not consist with moral nature, that there should be internal happiness or internal harmony, when the moral sense is made to suffer perpetual violence. A man of cunning and concealment, however dexterous, however triumphant in his worthless policy, is not at ease. The stoop, the downcast regards, the dark and sinister expression, of him who cannot lift up his head among his fellow men, or look his companions in the face, are the sensible proofs, that he who knows himself to be dishonest feels himself to be degraded; and the inward sense of dishonour which haunts and humbles him here, is but the commencement of that shame and everlasting contempt to which he shall awaken hereafter. This, you will observe, is a purely moral chastisement; and, apart altogether from the infliction of violence or pain on the sentient economy, is enough to overwhelm the spirit that is exercised thereby. Let him then that is unjust now be unjust still; and, in stepping from time to eternity, he bears, in his own distempered bosom, the materials of his coming vengeance along with him. The character itself will be the executioner of its own condemnation; and when, instead of each suffering apart, the unrighteous are congregated together—as in the parable of the tares, where, instead of each plant being severally destroyed, the order is given to bind them up in bundles and burn them—we may be well assured, that, where the turbulence and disorder of an unrighteous society are superadded to those sufferings which prey in secrecy and solitude within the heart of each individual member, a ten-fold fiercer and more intolerable agony will ensue from it. The anarchy of a state, when the authority of its government is for a time suspended, forms but a feeble representation of that everlasting anarchy, when the unrighteous of all ages are let loose to act and react with unmitigated violence on each other. In this conflict of assembled myriads; this fierce and fell collision between the outrages of injustice on the one side, and the outcries of resentment on the other; and, though no pain were inflicted, in this war of passions and of purposes, the passion and purpose of violence in one quarter calling forth the passion and the purpose of keenest vengeance back again—though no material or sentient agony were felt—though a war of disembodied spirits—yet in the wild tempest of emotions alone—the hatred, the fury, the burning recollection of injured rights, and the brooding thoughts of yet unfulfilled retaliation—in these, and these alone, do we behold the materials enough of a dire and dreadful pandemonium; and, apart from corporeal suffering altogether, may we behold, in the full and final developments of character alone, enough for imparting all its corrosion to the worm that dieth not, enough for sustaining in all its fierceness the fire that is not quenched.

But there is another moral ingredient in the future sufferings of the wicked, beside the one of which we have now spoken—suggested to us by the second clause of our text; and from which we learn that, not only will the unjust man carry his falsehoods and his frauds along with him to the place of condemnation, but that also the voluptuary will carry his unsanctified habits and unhallowed passions thitherward. "Let him that is filthy be filthy still." We would here take the opportunity of exposing, what we fear is a frequent delusion in society—who give their respect to the man of honour and integrity—and he does not forfeit that respect, though known at the same time to be a man of dissipation. Not that *we* think any one of the virtues, which enter into the composition of a perfect character, can suffer, without all the other virtues suffering along with it. We believe that a conjunction, between a habit of unlawful pleasure and the maintenance of a strict resolute exalted equity and truth, is very seldom, we could almost say, is never realised. The man of forbidden indulgence, in the prosecution of his objects, has a thousand degrading fears to encounter; and many concealments to practise; perhaps low and unworthy artifices to which he must descend; and how can either his honour or his humanity be said to survive, if at length, in his heedless and impetuous career, he shall trample on the dearest rights and the most sacred interests of families? With us it has all the authority of a moral aphorism, that the sobrieties of human virtue can never be invaded, without the equities of human virtue also being invaded. The moralities of human life are too closely linked and interwoven with each other, as that though one should be detached, the others might be left uninjured and entire; and so no one can cast his purity away from him, without a violence being done to the general moral structure and consistency of his whole character. But, be this as it may; we have the authority of the text and the oft reiterated affirmations of the New Testament, for saying of the voluptuary, that, if the countenance of the world be not withdrawn from him, the gate of heaven is at least shut against him; that nothing unclean or unholy can enter there; and that, carrying his uncrucified affections into the place of condemnation, he will find them too to be the ministers of wrath, the executioner of a still sorer vengeance. The loathing, the remorse, the felt and conscious degradation, the dreariness of heart that follow in the train of guilty indulgence here—these form but the beginning of his sorrows; and are but the presages and the precursors of that deeper wretchedness, which, by the unrepealed laws of moral nature, the same character will entail on its possessors in another state of existence. They are but the penalties of vice in embryo, and they may give at least the conception of what are these penalties in full. It will add—it will add inconceivably, to the darkness and disorder of that moral chaos, in which the impenitent shall spend their eternity—when the uproar of the bacchanalian and the licentious emotions is thus superadded, to the selfish and malignant passions of our nature; and when the frenzy of unsated desire, followed up by the languor and the compunction of its worthless indulgence, shall make up the sad history of many an unhappy spirit. We need not to dwell on the picture, though it brings out into bolder relief the all-important truth, that there is an inherent bitterness in sin; that, by the very constitution of our nature, moral evil is its own curse and its own worst punishment; that the wicked on the other side of death, but reap what they sow on this side of it; and that, whether we look to the tortures of a distempered spirit or to the countless ills of a distempered society, we may be very sure that to the character of its inmates—a character which they have fostered upon earth, and which now remains fixed on them through eternity—the main wretchedness of hell is owing.

Before quitting this part of the subject, we have but one remark more to offer. It may be felt as if we had overstated the power of mere character to beget a wretchedness at all approaching to the wretchedness of hell—seeing that the character is often realised in this world, without bringing along with it a distress or a discomfort which is at all intolerable. Neither the unjust man of our text, nor the licentious man of our text, is seen to be so unhappy here, in virtue of the moral characteristics which respectively belong to them, as to justify the imagination, that there, these characteristics will be of power, to effectuate such anguish and disorder of spirit as we have now been representing. But it is forgotten, first, that the world presents in its business, its amusements, and its various gratifications, a refuge from the mental agonies of reflection and remorse—and, secondly, that the governments of the world offer a restraint against the outbreakings of violence, which would keep up a perpetual anarchy in the species. Let us simply conceive of these two securities against our having even now a hell upon earth, that they are both taken down; that there is no longer such a world as ours, affording to each individual spirit innumerable diversions from the burden of its own thoughts; and no longer such a human government as ours, affording to general society a powerful defence against the countless variety of ills, that would otherwise rage and tumultuate within its borders—then, as sure as that a solitary prison is felt by every criminal to be the most dreadful of all punishments; and as sure as that, on the authority of law being suspended, the reign of terror would commence, and the unchained passions of humanity would go forth over the face of the land to raven and to destroy—so surely, out of moral elements and influences alone, might an eternity of utter wretchedness and despair be entailed on the rebellious: And, only let all the unjust and all the licentious of our text be formed into a community by themselves, and the Christianity which now acts as a purifying and preserving salt upon the earth be wholly removed from them; and then it will be seen that the picture has not been overcharged; but that the wretchedness is intense and universal, just because the wickedness reigns uncontrolled, without mixture and without mitigation.

But we now exchange this appalling for a delightful contemplation. The next clause of our text suggests to us the moral character of heaven. We learn from it that, on the universal principle "as a tree falleth so it lies," the righteous now will be righteous still. We no more dispute the material accompaniments of heaven, than we dispute the material accompaniments in the place of condemnation. But still we must affirm of the happiness that reigns, and holds unceasing jubilee there—that, mainly and pre-eminently, it is the happiness of virtue; that the joy of the eternal state is not so much a sensible or a tasteful or even an intellectual as it is a moral and spiritual joy; that it is a thing of mental, infinitely more than it is a thing of corporeal gratification; and, to convince us how much the former has the power and predominance over the latter, we bid you reflect, that, even in this world, with all the defect and disorder of its materialism, the curse upon its ground inflicting the necessity of sore labour, and the angry tempest from its sky after destroying or sweeping off the fruits of it, the infirmity of their feeble and distempered frames, after the pining sickness and at times the sore agony—yet, in spite of these, we ask whether it would not hold nearly if not universally true, that if all men were righteous then all men would be happy. Just imagine for a moment, that honour and integrity and benevolence were perfect and universal in the world; that each held the property and right and reputation of his neighbour to be dear to him as his own; that the suspicions and the jealousies and the heart-burnings, whether of hostile violence or envious competition, were altogether banished from human society; that the emotions, at all times delightful, of good-will on the one side, were ever and anon calling the emotions no less delightful of gratitude back again; that truth and tenderness hold their secure abode in every family; and, on stepping forth among the wider companionships of life, that each could confidently rejoice in every one he met with as a brother and a friend—we ask if on this simple change, a change you will observe in the *morale* of humanity, though winter should repeat its storms as heretofore, and every element of nature were to abide unaltered—yet, in virtue of a process and a revolution altogether mental, would not our millennium have begun, and a heaven on earth be realized? Now let this contemplation be borne aloft, as it were, to the upper sanctuary, where we are told there are the spirits of just men made perfect, or where those who were once the righteous on earth are righteous still. Let it be remembered, that nothing is admitted there, which worketh wickedness or maketh a lie; and that therefore, with every feculence of evil detached and dissevered from the mass, there is nought in heaven but the pure the transparent element of goodness—its unbounded love, its tried and unalterable faithfulness, its confiding sincerity. Think of the expressive designation given to it in the Bible, the land of uprightness. Above all think, that, revealed in visible glory, the righteous God, who loveth righteousness, there sitteth upon His throne, in the midst of a rejoicing family—Himself rejoicing over them, because, formed in His own likeness, they love what He loves, they rejoice in what He rejoices. There may be palms of triumph; there may be crowns of unfading lustre; there may be pavements of emerald, and rivers of pleasure, and groves of surpassing loveliness, and palaces of delight, and high arches in heaven which ring with sweetest melody—but, mainly and essentially, it is a moral glory which is lighted up there; it is virtue which blooms and is immortal there; it is the goodness by which the spirits of the holy are regulated here, it is this which forms the beatitude of eternity. The righteous now, who, when they die and rise again, shall be righteous still, have heaven already in their bosoms; and when they enter within its portals, they carry the very being and substance of its blessedness along with them—the character which is itself the whole of heaven's worth, the character which is the very essence of heaven's enjoyments.

"Let him that is holy, be holy still." The two clauses descriptive of the character in the place of celestial blessedness, are counterparts to the clauses descriptive of the character in the place of infernal woe. He that is righteous in the one stands contrasted with him that is unjust in the other. He that is holy in the one stands contrasted with him that is licentious in the other. But we would have you attend to the full extent and significance of the term "holy." It is not abstinence from the outward deeds of profligacy alone. It is not a mere recoil from impurity in action. It is a recoil from impurity in thought. It is that quick and sensitive delicacy to which even the very conception of evil is offensive—a virtue which has its residence within; which takes guardianship of the heart, as of a citadel or unviolated sanctuary in which no wrong or worthless imagination is permitted to dwell. It is not purity of action that is all which we contend for. It is exalted purity of sentiment—the ethereal purity of the third heavens, which, if once settled in the heart, brings the peace and the triumph and the unutterable serenity of heaven along with it. In the maintenance of this, there is a curious elevation; there is the complacency, we had almost said the pride, of a great moral victory over the infirmities of an earthly and accursed nature; there is a health and harmony to the soul; a beauty of holiness, which, though it effloresces on the countenance and the manner and the outward path, is itself so thoroughly internal, as to make purity of heart the most distinctive evidence of a work of grace in time, the most distinct and decisive evidence of a character that is ripening and expanding for the glories of eternity. "Blessed are the pure in heart, for they shall see God." "Without holiness no man shall see God." "Into the holy city nothing which defileth or worketh an abomination shall enter." These are distinct and decisive passages, and point to that consecrated way, through which alone, the gate of heaven can be opened to us. On this subject, there is a remarkable harmony, between the didactic sayings of various books in the New Testament, and the descriptive scenes which are laid before us in the book of Revelations. However partial and imperfect the glimpses there afforded of heaven may be, one thing is palpable as day, that holiness is its very atmosphere. It is the only element which its inmates breathe, and which it is their supreme and ineffable delight to breathe in. They luxuriate therein, as in their best-loved and most congenial element. Holiness is their oil of gladness—the elixir, if we may use the expression, the moral elixir of glorified spirits. And in their joyful hosannas, whether of "Holy, Holy, Holy, Lord God Almighty," or of "Just and true are thy ways thou King of Saints," we may read, that, as virtue in the Godhead is the theme of their adoration, so virtue in themselves is the very treasure they have laid up in heaven—the wealth, as well as the ornament, of their now celestial natures.

We would once more advert to a prevalent delusion that obtains in society. We are aware of nothing more ruinous, than the acquiescence of whole multitudes in a low standard of qualifications for Heaven. The distinct aim is to be righteous now, that, after the death and the resurrection, you may be righteous still—to be holy now, that you may be holy still. But hold it not enough, that you are free from the dishonesties which would forfeit the mere respect and confidence of the world, or from the profligacies which even the world itself would hold to be disgraceful. There is a certain amount of morality, which is in demand upon earth, but which is miserably short of the requisite preparation for Heaven—the holiness indispensable there, is a universal an unspotted and withal a mental and spiritual holiness. It is this which distinguishes the morality of a regenerated and aspiring saint, from the morality of a respectable citizen, who still is but a citizen of the world, with his conversation not in heaven, with neither his heart nor his treasure there. The righteous of our text would recoil from the least act of unfaithfulness, from being unfaithful in the least as from being unfaithful in much. The holy of our text would shrink in sensitive aversion and alarm from the first approaches of evil, from the incipient contaminations of thought and fancy and feeling, as from the foul and final contaminations of the outward history. Both are diligent to be found of Christ without spot and blameless, in the great day of account—glorifying the Lord with their soul and spirit, as well as with their bodies—aspiring after those graces, which, unseen by every earthly eye, belong to the hidden man of the heart, and in the sight of heaven are of great price—and so proceeding onward from strength to strength on this lofty path of obedience, till they appear perfect before God in Zion.

We feel that we have not nearly exhausted the subject of our text, by these brief and almost miscellaneous observations. The truth is, it is a great deal too unwieldy for any single address, and we shall therefore conclude with the notice of one specimen, that might be alleged for the importance of the view that we have just given, in purging theology from error. If the moral character then of these future states of existence, were distinctly understood and consistently applied, it would serve directly and decisively to extinguish antinomianism. It would in fact reduce that heresy to a contradiction in terms. There is no sound and scriptural Christian, who ever thinks of virtue as the price of heaven. It is something a great deal higher, it is heaven itself—the very essence, as we have already said, of heaven's blessedness. It occupies therefore a much higher place than the secondary and the subordinate one, ascribed to it even by many of the writers termed evangelical—who view it mainly as a token or an evidence that heaven will be ours. Instead of which it is the very substance of heaven—a sample on hand of the identical good, which, in larger measure and purer quality, is afterwards awaiting us—an entrance on the path which leads to heaven; or rather an actual lodgement of ourselves within that line of demarcation, which separates the heaven of the New Testament from the hell of the

New Testament. For heaven is not so much a locality as a character; and we, by a moral transition from the old to the new character, have in fact crossed the threshold, and are now rejoicing within the confines of God's spiritual family. By the doctrine of justification through faith, we understand that Christ purchased our right of admittance into heaven—or opened its door for us. Is there aught antinomian in this? The obstacle, the legal obstacle, between us and a life of prosperous and never-ending virtue, is now broken down; and is it upon that event, that we are to relinquish the path which has just been opened to welcome and invite our advancing footsteps? The doctrine of justification by faith is not an obstacle to virtue—it is but an introduction to it. It is in truth the removal of an obstacle—the unfastening of that drag which before held us in apathy and despair; and restrained us from breaking forth on that career of obedience, in which, with the hope of glory before us, we purify ourselves even as Christ is pure. The purpose of His death was not to supersede, but to stimulate our obedience. "He gave himself for us to redeem us from all iniquity and purify to himself a peculiar people zealous of good works." The object of His promises is not to lull our indolence, but rouse us to activity. "Having received these promises therefore, dearly beloved, let us cleanse ourselves from all filthiness of the flesh and spirit, perfecting holiness in the fear of God."

We expatiate no further; but shall be happy, f, as the fruit of these imperfect observations, you can be made to recognise how distinctly practical a business the work of Christianity is. It is simply to destroy one character, and to build up another in its room; to resist the temptations which vitiate and debase, and make all the graces and moralities which enter into the composition of perfect virtue the objects of our most strenuous cultivation. In the expediting of this mighty transformation, on the completion of which there hinges our eternity, we have need of believing prayer; a thorough renunciation of all dependence on our own strength; a thorough reliance on the proffered strength and aid of the upper sanctuary; a deep sense of our infirmities, and constant application for that Spirit who has promised to help them—that, in the language of the Apostle we may strive mightily, according to the grace which worketh in us mightily.

DISCOURSE VII.

ON THE REASONABLENESS OF FAITH.

"But before faith came, we were kept under the law, shut up unto the faith which should afterwards be revealed."—GALATIANS iii. 23.

"SHUT up unto the faith." This is the expression which we fix upon as the subject of our present discourse—and to let you more effectually into the meaning of it, it may be right to state, that in the preceding clause "kept under the law," the term *kept*, is, in the original Greek, derived from a word which signifies a sentinel. The mode of conception is altogether military. The law is made to act the part of a sentry, guarding every avenue but one—and that one leads those who are compelled to take it to the faith of the Gospel. They are shut up to this faith as their only alternative—like an enemy driven by the superior tactics of an opposing general, to take up the only position in which they can maintain themselves, or fly to the only town in which they can find a refuge or a security. This seems to have been a favourite style of argument with Paul, and the way in which he often carried on an intellectual warfare with the enemies of his Master's cause. It forms the basis of that masterly and decisive train of reasoning, which we have in his epistle to the Romans. By the operation of a skilful tactics, he, (if we may be allowed the expression) manœuvered them, and shut them up to the faith of the gospel. It gave prodigious effect to his argument, when he reasoned with them, as he often does, upon their own principles, and turned them into instruments of conviction against themselves. With the Jews he reasoned as a Jew. He made a full concession to them of the leading principles of Judaism—and this gave him possession of the vantage ground upon which these principles stood. He made use of the Jewish law as a sentinel to shut them out of every other refuge, and to shut them up to the refuge laid before them in the Gospel. He led them to Christ by a schoolmaster which they could not refuse—and the lesson of this schoolmaster, though a very decisive, was a very short one. "Cursed be he that continueth not in all the words of this law to do them." But, in point of fact, they had not done them. To them then belonged the curse of the violated law. The awful severity of its sanctions was upon them. They found the faith and the free offer of the Gospel to be the only avenue open to receive them. They were shut up unto this avenue; and the law, by concluding them all to be under sin, left them no other outlet but the free act of grace and of mercy laid before us in the New Testament.

But this is not the only example of that peculiar way in which St. Paul has managed his discussions with the enemies of the faith. He carried the principle of being all things to all men into his very reasonings. He had Gentiles as well as Jews to contend with—and he often made some sentiment or conviction of their own, the starting point of his argument. In this same epistle to the Romans, he pleaded with the Gentiles the acknowledged law of nature and of conscience. In his speech to the men of Athens, he dated his argument from a point in their own superstition. In this way he drew converts both from the ranks of Judaism, and the ranks of idolatry—and whether it was the school of Gamaliel in Jerusalem, or the school of poetry and philosophy in countries of refinement, that he had to contend with, his accomplished mind was never at a loss for principles by which he bore down the hostility of his adversaries, and shut them up unto the faith.

But there is a fashion in philosophy as well as in other things. In the course of centuries, new schools are formed; and the old, with all their doctrines, and all their plausibilities, sink into oblivion. The restless appetite of the human mind for speculation, must have novelties to feed upon—and after the countless fluctuations of two thousand years, the age in which we live has its own taste, and its own style of sentiment to characterize it. If Paul, vested with a new apostolical commission, were to make his appearance amongst us, we should like to know how he would shape his argument to the reigning taste and philosophy of the times. We should like to confront him with the literati of the day, and hear him lift his intrepid voice in our halls and colleges. In his speech to the men of Athens, he refers to certain of their own poets. We should like to hear his references to the poetry and the publications of modern Europe—and while the science of this cultivated age stood to listen in all the pride of academic dignity, we should like to know the arguments of him who was determined to know nothing save Jesus Christ, and Him crucified.

But all this is little better than the indulgence of a dream. St. Paul has already fought the good fight, and his course is finished. The battles of the faith are now in other hands—and though the wisdom, and the eloquence, and the inspiration of Paul have departed from among us, yet he has left behind him the record of his principles. With this for our guide, we may attempt to do what he himself calls upon us to do. We may attempt to be followers of him. We may imitate him in the intrepid avowal of his principles—and we may try, however humbly and imperfectly, to imitate his style of defending them. We may accommodate our argument to the reigning principles of the day. We may be all things to all men—and out of the leading varieties of taste and of sentiment which obtain in the present age, and in the present country, we may try if we can collect something, which may be turned into an instrument of conviction for reclaiming men from their delusions, and shutting them up unto the faith.

There is first, then, the school of Natural Religion—a school founded on the competency of the human mind to know God by the exercise of its own faculties—to clothe Him in the attributes of its own demonstration—to serve Him by a worship and a law of its own discovery—and to assign to Him a mode of procedure in the administration of this vast universe, upon the strength and the plausibility of its own theories. We have not time at present for exposing the rash and unphilosophical audacity of all these presumptions. We lay hold of one of them; and we maintain, that if steadily adhered to, and consistently carried into its consequences, it would empty the school of natural religion of all its disciples—it would shut them up unto the faith, and impress one rapid and universal movement into the school of Christ. The principle which we allude to makes a capital figure in their self-formed speculations; and it is neither more nor less than the judicial government of God over moral and accountable creatures. They hold that there is a law. They hold the human race to be bound to obedience. They hold the authority of the law to be supported by sanctions; and that the truth, and justice, and dignity of the supreme Being are involved in these sanctions being enforced and executed. One step more, and they are fairly shut up unto the faith. That law which they hold to be in full authority and operation over us, has been most unquestionably violated. We appeal, as Paul did before us, to the actual state of the human heart, and of human performances. We ask them to open their eyes to the world around them—to respect, like true philosophers, the evidence of observation, and not to flinch from the decisive undeniable fact which this evidence lays before them. Men are under the law, and that law they have violated. "There is not a just man on earth, that sinneth not." It is not to open shameless and abandoned profligacy, that we are pointing your attention. We make our confident appeal to the purest and loveliest of the species. We rest our cause with the most virtuous individual of our nature. We enter his heart, and from what passes there, we can gather enough, and more than enough, to overthrow this tottering and unsupported fabric. We take a survey of its desires, its wishes, its affections—and we put the question to the consciousness of its possessor, if all these move in obedient harmony even to the law of natural religion. The external conduct viewed separately and in itself, is, in the eye of every enlightened moralist, nothing. It is mere visible display. Virtue consists in the motive which lies behind it; and the soul is the place of its essential residence. Bring the soul then into immediate comparison with the law of God. Think of the pure and spiritual service which it exacts from you. Amid all the busy and complicated movements of the inner man, is there no estrangement from God? Are there no tumultuous wanderings from that purity, and goodness, and truth, which even philosophers ascribe to Him? Is there no shortcoming from the holiness of His law, and the magnificence of His eternity? Is there no slavish devotion to the paltry things of sense and of the world? Is there no dreary interval of hours together, when God is unfelt and unthought of? Is there no one time when the mind delivers itself up to the guidance of its own feelings, and its own vanities—when it moves at a distance from heaven—and, whether in solitude or among acquaintances, carries along, without any reference to that Being whose arm is perpetually upon me; who, at this moment, is at my right hand, and measures out to me every hairbreadth of my existence—who upholds me through every point of that time which runs from the first cry of my infancy, to that dark hour when the weight of my dying agonies is upon me—whose love and whose kindness are ever present, to give me every breath which I draw, and every comfort which I enjoy? We grant the disciples of natural religion the truth of their own principle, that we are under the moral government of the Almighty —and by the simple addition of one undeniable fact to their speculation, we shut them up unto the faith. The simple fact is, that we are rebels to that government; and the punishment of these rebels is due to the vindication of its insulted authority To say, that God will perpetually interpose with an act of oblivion, would be vastly convenient for us—but what then becomes of that moral government which figures away in the demonstrations of moralists? Does it turn out after all, to be nothing more than an idle and unmeaning declamation, on which they love to expatiate—without any thing like real attention or belief on the part of the thinking principle? If they are not true to their own professed convictions, we can undertake to shut them up to nothing. This is slipping from under us—but it is by an actual desertion of their own principle. If you cannot get them to stand to the argument, the argument is discharged upon them in vain. If this be the result, we do not promise ourselves that all we can say shall have any weight upon their convictions—not, however, because they have gained a victory, but because they have betaken themselves to flight. At the very moment that we thought of shutting them up, and binding them in captivity to the obedience of the truth, they have turned about and got away from us—but how? By an open renunciation of their own principle. Look at the great majority of infidel and demi-infidel authors, and they concur in representing man as an accountable subject, and God as a judge and a lawgiver. Examine then the account which this subject has to render—and you will see, in characters too glaring to be resisted, that with the purest and most perfect individual amongst us, it is a wretched account of guilt and deficiency. What make you of this? Is the subject to rebel and disobey every hour, and the King, by a perpetual act of indulgence, to efface every character of truth and dignity from His government? Do this, and you depose the legislator from His throne. You reduce the sanctions of His law to a name and a mockery. You give the lie to your own speculation. You pull the fabric of His moral government to pieces--and you give a spectacle to angels which makes them weep compassion on your vanity—poor, pigmy, perishable man, prescribing a way to the Eternal, and bringing down the high economy of Heaven to the standard of his convenience, and his wishes. This will never do. If there be any truth in the law of God over the creatures whom He has formed, and if that law we have trampled upon, we are amenable to its sentence. Ours is the dark and unsheltered state of condemnation—and if there be a single outlet or way of escaping, it cannot be such a way as will abolish the law, and degrade the Lawgiver—but it must be such a way as will vindicate and exalt the Deity—as will pour a tide of splendour over the majesty of His high attributes—and as in the sublime language of the prophet, who saw it from afar, will magnify His law, and make it honourable. To this way we are fairly shut up. It is our only alternative. It is offered to us in the Gospel of the New Testament. I am the way, says the Author of that Gospel, and by me, if any man enter in, he shall be saved. In the appointment of this Mediator—in His death, to make propitiation for the sins of the world—in His triumph over the powers of darkness—in the voice heard from the clouds of heaven, and issuing from the mouth of God himself, "This is my beloved Son, in whom I am well pleased"—in the resistless argument of the Apostle, who declares God to be just, and the justifier of him that believeth in Jesus—in the undoubted miracles which accompanied the preaching of this illustrious personage, and His immediate followers—in the noble train of prophecy, of which He was the object and the termination—in the choir of angels from heaven, who sung His entrance into the world—and in the sublime ascension from the grave, which carried him away from it—in all this we see a warrant and a security given to the work of our redemption in the New Testament, before which philosophy and all her speculations vanish into nothing. Let us betake ourselves to this way. Let us rejoice in being shut up unto it. It is passing, in fact, from death unto life—or, from our being under the law, which speaks tribulation and wrath to every soul of man that doeth evil, to being under the grace which speaks quietness and assurance for ever to all that repair to it. The scripture hath concluded all to be under sin, that the promise by faith of Jesus Christ might be given to them that believe.

We now pass on from the school of natural religion to another school, possessing distinct features —and of which we conceive the most expressive designation to be, the school of Classical Morality. The lessons of this school are given to the public in the form of periodical essays, elaborate dissertations on the principles of virtue, eloquent and often highly interesting pictures of its loveliness and dignity, the charm that it imparts to domestic retirement, and its happy subservience to the peace, and order, and well-being of society. It differs from the former school in one leading particular. It does not carry in its speculations so distinct and positive a reference to the Supreme Being. It is true, that our duties to Him are found to occupy a place in the catalogue of its virtues; but then the principle on which they are made to rest, is not the will of God, or obedience to His law. They are rather viewed as a species of moral accomplishment —the effect of which is to exalt and embellish the individual. They form a component part of what they call virtue—but if virtue be looked upon in no other light than as the dress of the mind, we maintain, that in the act of admiring this dress, and of even attempting to put it on, you may stand at as great a distance from God, and He be as little in your thoughts, as in the tasteful choice of your apparel, for the dress and ornament of the body. The object of these writers is not to bring their readers under a sense of the dominion and authority of God. The main principle of their morality, is not to please God, but to adorn man—to throw the splendour of virtue and accomplishment around him—to bring him up to what they call the end and the dignity of his being—to raise him to the perfection of his nature—and to rear a spectacle for the admiration of men and of angels, whom they figure to look down with rapture, from their high eminence, on the perseverance of a mortal in the career of worth, and integrity, and honour. This is all very fine. It makes a good picture—but what we insist upon is, that it is a fancy picture; that, without the limits of Christianity and its influence, you will not meet with a single family, or a single individual, to realize it—that the whole range of human experience furnishes no resemblance to it—and that it is as unlike to what we find among the men of the world, or in the familiar walks of society, as the garden of Eden is unlike the desolation of a pestilence. The representation is beautiful—but more flattering than it is fair. It is a gaudy deception, and stands at as great a distance from the truth of observation, as it does from the truth of the New Testament. There is positively nothing like it in the whole round of human experience. It is the mere glitter of imagination. It may serve to throw a tinsel colouring over the pages of an ambitious eloquence—but with business and reality for our objects, we may describe the [illegible] of many thousand families, or take our station for years in the market-place, and in our attempts to realize the picture which has been laid before us, we will be sure to meet with nothing but vanity, fatigue, and disappointment. Now, the question we have to put to the disciples of this school is, are they really sincere in this admiration of virtue? Is it a true process of sentiment within them? We are willing to share in their admiration, and to ascend the highest summit of moral excellence along with them. We join issue with them on their own principle, and coupling it with the obvious and undeniable fact of man's depravity, we shut them up unto the faith. Virtue is the idol which they profess to venerate—and this virtue, as it exists in their own conceptions, and figures in their own dissertations, they cannot find. In proportion to their regard for virtue, must be their disappointment at missing her—and when we witness the ardour of their sentiments, and survey the elegance of their high-wrought pictures, what must be the humiliation of these men, we think, when they look on the world around them, and contrast the purity of their own sketches, with the vices and the degradation of the species. Grosser beings may be satisfied with the average morality of mankind—but if there be any truth in their high standard of perfection, or any sincerity in their aspirations after it, it is impossible that they can be satisfied. By one single step do we lead them from the high tone of academic sentiment, to the sober humility of the Gospel. Give them their time to expatiate on virtue, and they cannot be too loud or eloquent in her praises. We have only a single sentence to add to their description: The picture is beautiful, but on the whole surface of the world we defy them to fasten upon one exemplification—and by every grace which they have thrown around their idol, and every addition they have made to her loveliness, they have only thrown mankind at a distance more helpless and more irrecoverable from their high standard of duty and of excellence.

The tasteful admirer of eloquent description and beautiful morality, turns with disgust from those mortifying pictures of man, which abound in the New Testament. We only ask them to combine, with all this finery and eloquence, what has been esteemed as the best attribute of a philosopher, respect for the evidence of observation. We ask them to look at man as he is, and compare him with man as they would have him to be. If they find that he falls miserably short of their ideal standard of excellence, what is this but making a principle of their own the instrument of shutting them up unto the faith of the Gospel, or, at least, shutting them up unto one of the most peculiar of its doctrines, the depravity of our nature, or the dismal ravage which the power of sin has made upon the moral constitution of the species? The doctrine of the academic moralist, so far from reaching a wound to the doctrine of the Apostle, gives an additional energy to all his sentiments. "My mind approves the things which are more excellent, but how to perform that which is good, I find not." "I delight in the law of God after the inward man." "But the good that I would I do not, and the evil that I would not, that I do."

But the faith of the Gospel does not stop here. It does not rest satisfied with shutting us up unto a belief of the fact of human depravity. That depravity it proposes to do away. It professes itself equal to the mighty achievement of rooting out the deeply seated corruption of our nature—of making us new creatures in Christ Jesus—of destroying the old man and his deeds, and bringing every rebellious movement within us under the dominion of a new and a better principle. If sincere in your admiration of virtue, you are shut up unto the only expedient for the re-establishment of virtue in the world. That expedient is the Spirit of God working in the heart of believers—quickening those who were dead in trespasses and sins, and bringing into action the same mighty power which raised Jesus from the grave, for raising us who believe in Jesus to newness of life and of obedience. This is the process of sanctification laid before us in the New Testament. A wonderful process it undoubtedly is—but are we who walk in a world of mystery, who have had only a few little years to look about us, and are bewildered at every step amid the variety of God's works and of His counsels, are we to reject a process because it is wonderful? Must no step, no operation of the mighty God be admitted, till it is brought under the dominion of our faculties?—and shall we who strut our little hour in the humblest of His mansions, prescribe a law to Him whose arm is abroad upon all worlds, and whose eye can take in, at a single glance, the unmeasurable fields of creation and providence? Be it as wonderful as it may—enough for us that it is made sure by the distinct and authentic testimony of heaven—and if, from the mouth of Jesus, who is heaven's messenger, we are told, that "unless a man be born of the Spirit, he cannot enter into the kingdom," it is our part submissively to acquiesce, and humbly to pray for it. Whatever repugnance others may feel to this part of the revealed counsels of God, those who look to a sublime standard of moral excellence, and sigh for the establishment of its authority in the world, ought to rejoice in it. It is the only remaining expedient for giving effect and reality to their own declamations, and they are fairly shut up unto it. Long have they tried to repair the disorders of a ruined world. Many an expedient has been fallen upon. Temples have been reared to science and to virtue—and from the lofty academic chair, the wisdom of this world has lifted its voice amid a crowd of listening admirers. For thousands of years, the unaided powers and principles of humanity, have done their uttermost—and tell us, ye advocates for the dignity of the species, the amount of their operation. If you refuse to answer, we shall answer for you—and do not hesitate to say, that mighty in promise, and wretched in accomplishment, you have positively done nothing—that all the wisdom of the schools, and all its vapouring demonstrations, have not had the least perceptible weight, when brought to bear upon the mass of human character, and human performance—that the corruption of the inner man has not yielded at all to your reasoning, and remains as unsubdued and as obstinate a principle as ever—that the power of depravity in the soul of man is beyond you—and that setting aside the real operation of Christianity in the hearts of individuals, and the surface dressing which the hand of legislation has thrown over the face of society, the human soul, if seen in its nakedness, would still be seen in all its original deformity—as strong in selfishness, as lawless in propensity, as devoted to sense and to time, as estranged from God, as unmindful of the obedience, and as indifferent to the reward and the inheritance of His children.

The machine has gone into disorder—and there is not a single power within the compass of the machinery itself that is able to repair it. You must do as you do in other cases—you must have recourse to some external application. The inefficacy of every tried expedient shuts you up unto the only remaining one. Every human principle has been brought to bear upon it in vain; and we are shut up unto the necessity of some other principle that is beyond humanity, and above it. The Spirit of God is that mighty principle. That Spirit which moved on the face of the waters, and made light, and peace, and beauty to emerge out of the wild war of nature and her elements, is the revealed agent of Heaven, for repairing the disorders of sin, and restoring the moral creation of God to health and to loveliness. It will create us anew unto good works. It will make us again after that image in which we were originally formed. It will sanctify us by the faith that is in Jesus. And by that mighty power whereby it is able to subdue all things unto itself, it will obtain the victory over that spirit which now worketh in the children of disobedience. The resurrection of Jesus from the dead is the first fruit of its operation—and to him who believes it is the satisfying pledge of its future triumphs. That body, which, left to itself, would have mouldered into fragments, is now in all the bloom of immortality, at the right hand of the everlasting throne. We have tried the operation of a thousand principles in vain. Let us repair to this, so great in promise, and so mighty in performance. It has already achieved its wonders. It has wrought those miracles of faith and fortitude which, in the first ages of Christianity, threw a gleam of triumph over the horrors of martyrdom. It has given us displays of the great and the noble which are without example in history—and from the first moment of its operation in the world, it has been working in those unseen retirements of the cottage and the family, where the eye of the historian never penetrates. The admirers of virtue are fairly shut up unto the faith—for faith is the only avenue that leads to it. "To your faith add virtue," says the Apostle—and that you may be able to make the addition, the promise of the Spirit is given to them that believe.

We should now pass on to another school, the school of fine feeling and poetical sentiment. It differs from the former in this—that while the one, in its dissertations on virtue, carries us up to the principles of duty, the other paints and admires it as a tasteful exhibition of what is fair and lovely in human character. The one makes virtue its idol because of its rectitude; the other makes virtue its idol because of its beauty—and the process of reasoning by which they are shut up unto the faith, is the same in both. Look at the actual state of the world, and we find that both the rectitude and the beauty are a-wanting. If you admire the one, and love the other, you are shut up unto the only expedient that is able to restore them—and that expedient is sanctioned by the truth of heaven, and has all the power of omnipotence employed in giving effect to the operation—the Spirit of God subduing all things unto itself—putting the law in our hearts, and writing it in our minds—and by bringing the soul of man under the influence of "whatsoever things are pure, or honest, or lovely, or of good report," creating a finer spectacle, and rearing a fairer and more unfading flower, than ever grew in the gardens of poetry.

The processes are so entirely similar, that we would not have made it the distinct object of your attention, had it not been for the sake of an argument in behalf of the faith, which may be addressed with great advantage to the literary and cultivated orders of society. There are few people of literary cultivation, who have not read a novel. In this fictitious composition, there are often one or two perfect characters that figure in the history, and delight the imagination of the reader—and you are at last landed in some fairy scene of happiness and virtue, which it is quite charming to contemplate, and which you would like to aspire after—perhaps some interesting family in the bosom of which love, and innocence, and tranquillity, have fixed themselves—where the dark and angry passions never enter—where suspicion is unknown, and every eye meets another in the full glance of cordiality and affection—where charity reigns triumphant, and smiles beneficence and joy upon the humble cottages which surround it. Now this is very soothing, and very delightful. It makes one glad to think of it. The fancy swells with rapture, and the moral principle of our nature lends its full approbation to a scene so virtuous and so exemplary. So much for the dream of fancy. Let us compare it with the waking images of truth. Walk from Dan to Beersheba; and tell us, if, without and beyond the operation of Gospel motives and Gospel principles, the reality of life ever furnished a picture that is at all like the elegance and perfection of this fictitious history. Go to the finest specimen of such a family. Take your secret stand, and observe them in their more retired and invisible moments. It is not enough to pay them a ceremonious visit, and observe them in the put on manners and holiday dress of general company. Look at them when all this disguise and finery are thrown aside. Yes, we have no doubt, that you will perceive some love, some tenderness, some virtue—but the rough and untutored honesty of truth compels us to say, that along with all this, there are at times mingled the bitterness of invective, the growlings of discontent, the harpings of peevishness and animosity, and all that train of angry, suspicious, and discordant feelings, which embitter the heart of man, and make the reality of human life a very sober affair indeed, when compared with the high colouring of romance, and the sentimental extravagance of poetry.

Now, what do we make of all this? We infer, that however much we may love perfection, and aspire after it, yet there is some want, some disease in the constitution of man, which prevents his attainment of it—that there is a feebleness of principle about him—that the energy of his practice does not correspond to the fair promises of his fancy—and however much he may delight in an ideal scene of virtue and moral excellence, there is some lurking malignity in his constitution, which, without the operation of that mighty power revealed to us in the Gospel, makes it vain to wish, and hopeless to aspire after it.

END OF VOLUME SEVENTH.

www.ingramcontent.com/pod-product-compliance
Lightning Source LLC
LaVergne TN
LVHW010153110826
845151LV00002B/503